To my wife Linda, who knows the back of my head really, really well -

Thank you.

(*Now* do you still love me?)

William,
Best of luck!

Still MORE Acknowledgements...

My thanks to Al, Ian, & others at SmartHomeUSA.com for their feedback & encouragement. Also to Jeff at JDS for giving me something to write about in the "headaches" section (seriously, he was helpful).

I also need to thank everyone for the use of some of the photos/illustrations, though the vast majority were my own. That's why some are blurry...

Part One:

You Can Do This!

Home automation is sort of a black art. No, there's nothing evil or immoral about it, but there just aren't any rules. There's no industry standard that determines when you've "automated" your home. Some who do this for a living will plug your lamps into some X-10 modules, give you an RF keychain remote, and pronounce your house "smart."

On the other extreme, some really sophisticated commercial systems are occasionally used in upscale homes where a lot of money exchanges hands, but the system is never used to its full potential... or then again maybe it is, depending on the skill level and work ethic of the installer.

The average person has little way of knowing. Add to this the many different products all claiming to do virtually the same thing - and you have confusion for the consumer.

In this book I plan to help remove that confusion and give you the tools to at least 1.) make informed decisions, but *preferably* 2.) get the proper equipment and infrastructure in place (if you want to do this yourself) for a home that will truly amaze you.

You may very well be the first on the block with a home that knows how to behave itself!

Having said that, there are two things I don't want to do in this book. One is to regale you with a lot of personal stories and filler material (which only makes the book look bigger); and the other is to be so dry and technical that you miss the point of how simple and fun accomplishing your automation goals can be (I mean, there are REASONS for doing this beyond simply proving that you can).

- **I haven't forgotten to take the trash out in years!**

- **My son takes his dirty shoes off at the door - *and I don't have to say a word to him!***

- **I can password-protect outgoing calls to *900* numbers, directory assistance, long-distance or any other specific phone number.**

- **If I want, I can receive phone calls from home notifying me of just about anything that occurs: someone arrived home, water in the basement, messages in voicemail - *just about anything!***

- **It *always* looks like someone's home at my house.**

My Approach...

There are a number of "smart controllers" available on the market. If you're into Crestron or Phast, then this book probably isn't for you. I'm not taking you down the road of high dollar stuff. Besides, if that's what you want to use, you can probably afford to hire a professional anyway *(but you'll be missing out on the fun, in my opinion).*

Here, I've decided to use what I believe gives the most bang for the buck, the JDS Stargate. Along the way, I'll also point out how to accomplish many of the same objectives without the expense of a centralized "brain" like Stargate.

That way, whether you decide to purchase one for

yourself or not, you'll still be equipped to do what you want within your own budget. And later, if you decide to upgrade - no problem! What you do now can easily be pulled together later.

What we're going to do is start from the assumption that you know very little (but give you credit for having a brain), and show you step-by-step how you can decide just how much time and money you want to invest, as well as deciding for yourself how much sophistication you need.

The really neat thing about the approach we're going to take is that you don't have to have all the answers to these questions before you start. To a large degree, you can just dive in and start doing some cool things without worrying about whether what you're doing right now is going to fit into your final design!

The reason for this is that I'm going to take you through a MODULAR approach. You can build upon anything you do and INTEGRATE it into your future plans if you follow a few simple guidelines:

- **Don't** use any proprietary hardware
- **Don't** use any proprietary protocols
- ***DO*** use an "open architecture"

I'll get into this in more detail shortly, but what it simply means is that there are plenty of protocols & methods which are common across the industry. If you stick to these, you'll have no trouble INTEGRATING different pieces of hardware.

But if you use something in your house that only speaks its own "language," you may later regret it.

So let's move on. The first couple of sections may seem a little wordy (I know - promised I wouldn't do that), but I really think some introductory material is important - just so you know where I come from & why I recommend the approach that I do . . . .

How NOT to Waste a Lot of Money

I'm currently a network engineer (MCSE, CCNA, CCNP), but prior to this I ran my own audio/video business for several years.

Over time, I became aware that more of my customers were turning into do-it-yourselfers, using me to wire their homes and then counting on me to advise them how to finish.

While they didn't always do it "right" (by my standards), nor did they always pick the best products for their money, they seemed to take satisfaction in the ownership of their work.

Of course, there were the high-end customers who really didn't want to take a chance on messing up an expensive installation, and preferred to pay me to take that chance for them.

The problem that became apparent, though, was that the more upscale the theater room was, the more difficult it was to operate.

It wasn't too hard to get lost even if you knew what you were doing, and woe to the customer if he was sitting in the dark fumbling with the remoteS (no, that's not a typo) trying to figure out what mode he was in, whether his audio was digital or analogue,

. . . *and why* ***in the world*** he was listening to classical music while watching Arnold and the Predator go at it.

This was what initially got me interested in automation. It would be great, I thought, if you only had to figure out how to operate everything ONCE, store the various sequences in memory, and call upon those sequences with a single command.

Eventually, as a logical progression my interest turned to whole house automation. Again I thought, wouldn't it be great if your house knew whether you were at home or not, and behaved accordingly? And if it was dark, it would behave one way - if light, another?

And if you were on vacation, wouldn't only the stupidest burglar want to test your home if he was convinced that it was patrolled by a roving gang of Dobermans? I mean, even if he did try, the evidence would be there on your hidden VCR - or PC - or better, remote Web Site - that he so much as entered your yard - I mean, wouldn't that be great?

It seemed obvious to me; but actually, one of the difficulties I've faced in this business is communicating the advantages of integrating home electronics.

Usually I meet a customer who is initially planning to have a theater room or distributed audio system installed. When I start talking about automating other features of the home the first response is typically a blank stare.

I think it's probably because I live in the Midwest. Here, most people either don't understand the practicality (not to mention the fun), or else they've heard the horror stories of installations that simply didn't work. It then takes a little enthusiasm on my part to convince people that this is worth doing.

But, oh man, what's the best way to make it all happen?

There have, of course, been expensive systems available for a long time. They work great, and look great - and cost a lot.

I mean **A LOT.**

So I began researching a number of them, looking for that perfect device which would fit into my own meager budget, and still accomplish everything I needed (after all, I reasoned, I couldn't sell something like this that I hadn't personally owned and played with).

Before you invest money in any equipment that has "intelligent" capabilities, whether it's a security system, remote controlled ceiling fans, smart thermostats - think about what it is that you want to accomplish. You can spend a lot of money on numerous gadgets, or you can spend a moderate amount on ONE smart controller which will replace the capabilities of all the others. JDS Stargate is one controller that I highly recommend for its flexibility and ease of use.

I can now say that I have personally owned three systems, and the most expensive of the three is the one that is actually the least useful. It costs two to three times as much as the Stargate, and is utterly inflexible in terms of what it will do. It's designed for a certain amount of glitz, and in my opinion that's about it.

To be fair, there are a lot of people (famous people) who have this particular system in their homes and are very happy with it, but in my opinion that may well be because they don't know what they're missing. I attended dealer training for this product, and when I asked about certain capabilities, I was simply told "not as intended"

- which was interpreted by me as "it can't **do** that."

The obvious point I'm making is that you can easily waste a lot of money.

Well, fear not, for my objective is to help you in that area (especially since you're doing this yourself and I'm not making anything from you anyway - aside from selling this book). The research I've done. The experience I have. I probably have all I'm gonna get from you.

Some Basic Terminology:

I'm starting from the assumption that this is something that already interests you (that's why you're reading this, right?).

And so to begin with, let's get some terminology straight:

HOME AUTOMATION- This is actually a bad word, despite the fact that I've been using it. It's pretty nebulous, & could mean just about anything. I use it just because it's what so many other people use.
. . . But a better word is System Integration.

SYSTEM INTEGRATION- Good word. It has meaning. It refers to the art of taking the electronic systems in the home and bringing them all under central control so that they work together in an efficient and effective manner. Heating & Air Conditioning, Lights, Stereo, TV's, VCR's, Security System, Cameras, Doorbells, Motion Detectors - all work together to make life at home more secure, economical, and comfortable.

While plugging in an X-10 module may give you remote (or even timer-based) control, it is only *one* part of an integrated home.

PROPRIETARY VS. OPEN ARCHITECTURE- In the past there were some systems that came out on the market promoting themselves as the way of the future. The problem that arose down the road was that those manufacturers either decided to stop making and supporting those products, or they simply went out of business.

Now, this wouldn't be such a terrible thing for the end user except for the fact that some of these systems used proprietary hardware and protocols. That meant that the "brain" could only communicate with hardware of its own unique kind, and the wiring infrastructure was suitable for only its own system; so when the product was no longer available, replacement parts and upgrades became impossible to get.

Needless to say, I do *not* recommend a proprietary system.

An open architecture is much to be preferred. It is scaleable, upgradeable, works with many, many products from other manufacturers, and uses protocols and wiring schemes that are extremely common, such as X-10, low voltage inputs and outputs, relays, ASCII code (computer talk), DTMF (touch-tone telephones), infrared, and UHF (radio frequency).

Nothing mysterious about these, but they are sufficient to do virtually anything that can be done! Plus, there is an abundance of interchangeable accessory devices available, and any prewiring that you might do is common to the open architecture of them all.

MANUAL CONTROL: Well now, what does that sound like? The truth is, if you can't turn on your lights by hitting a light switch, then you've lost control (as well as your family's good will).

A must in any good design is the ability for the homeowner to assume direct control over everything if he wants to do so. As someone else said, "if grandma comes for a visit, everything should work just like she expects." It's poor planning when automation becomes the ONLY way for anything to work.

Maybe it's becoming obvious by now that there are certain things I take for granted.

- First, you should get your money's worth.

- Second, your design should have the inherent capability to do just about anything imaginable (or to be upgradeable to do so).

- Third, YOU as the homeowner should be able to control the system, or at least to override it manually.

- And fourth, you should love this stuff as much as I do.

You really should.

What Kind of Stuff to Use?

Alright, now we begin to take a closer look at some of the common protocols that I mentioned earlier. Don't worry about understanding everything in detail, but just familiarize yourself with the following and remember that you can refer back to these pages later if need be.

Also, even if some of what I'm about to say seems to insult your intelligence (after all, I don't know how much you know), don't think that things don't get pretty intricate later on. The "basics" can be put together in some pretty unique ways, as you'll soon see.

INFRARED:

Consider your television for a moment - or your VCR - or your stereo. They utilize one of the most common forms of communication: infrared (IR). In case you're still turning on the TV by pulling the knob, IR is an invisible spectrum of light which is used to communicate different commands to different components by varying the frequencies and pulse counts. It's a great way to communicate without having direct hardwired control.

However, some of the problems associated with it are its limited range and need for direct line-of-sight, meaning it won't work through walls or people, etc. There are a couple of means to overcome these hindrances. As a for-instance, let's say you're watching your satellite dish on your bedroom TV; but your problem is that the satellite receiver itself is located in the family room along with the main TV, and that is where you need to be in order to change channels (remember direct line-of-sight?).

Your choice of solutions in this case may be a common one: an IR Extender device, such as the RCA Powermid. This consists of two cone-shaped devices, Cone #1 located next to the satellite receiver, and the other (Cone #2) in the remote bedroom. Pointing your remote at Cone #2, you issue a command to change channels. Cone #2 has no problem receiving your command, because it's plainly visible to your remote. However, Cone #1 is several rooms away, and it needs to get that same information in order to relay it to the satellite receiver.

What happens in this case is that the IR is transformed to UHF signals by Cone #2, and then sent right through the walls and the bodies of anybody standing in the way to Cone #1, which then converts it back to IR and re-issues the command. This is an inexpensive and quick way to get IR from one room to another, but it's not the best. However, if it works for you there's no reason not to be happy.

In any quality installation that I do, I use a hardwired IR system, which I'll discuss more in a little while. Though it costs more, the advantages of this kind of installation are substantial. It works better over longer distances than the Extender type device, and in my opinion has more of a custom "permanent" feel to it.... Well, at any rate, it just plain works better.

UHF:

We just touched on an application of UHF. Radio frequency is not hindered by obstructions to the degree that IR is. However, if enough stuff gets in the way, you'll have problems. Besides, there aren't enough purely UHF devices being made for it to be much of an issue, except for maybe a "premium" UHF satellite system (by the way, you might take note: a UHF satellite receiver presents too much of an automation problem unless you get into more expensive automation designs - if you're shopping satellite I'd recommend that you get an IR receiver).

Give me hardwired IR over UHF any day.

X10:

This is something that's been around for a long time. It's fought long and hard to earn some of its improvements in reliability, and has a definite place in your home. A real purist may object to some of the potential problems with it - but unless money is no object to you, you can't beat the affordability and practical flexibility of X10. Okay, so "what is it" you might ask? Another name for it is "powerline carrier technology," which is really a more descriptive term, because that's exactly how the signal is transmitted. Signal actually travels the sine wave of household AC current; and in keeping with that, let it be said that X10 theory abounds. But I'm not going to bore you with more than you need to know - and here it is:

X10 code can be sent to 256 different addresses, made up of varying combinations of House and Unit Codes. There are sixteen House Codes, ranging alphabetically from A to P. Each of these House Codes can in turn be numbered from "one" to "sixteen" (Unit Codes), giving you 256 different possible addresses (the ability to control 256 different devices).

A simple but common application would be to turn on lights with an X10 remote, say, from your car as you pull into your driveway. This is so easy to do you'll have this under your belt in the next 60 seconds. Get an X10 UHF Receiver (base) module, set its address to A-1 (or whatever) and plug it into a wall outlet, and then plug your lamp into the module (make sure the lamp is in the "on" position. Now, with your X10 remote (set to the same address) you can turn the light on/off from probably 20 - 50 feet away. It uses RF (radio frequency), so you won't need line of sight like infrared.

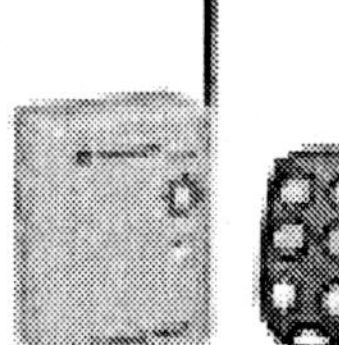 However, maybe for security's sake you need the light to come on at dusk and remain on until dawn. There are X10 controllers that operate by a timer or by photocell. You can buy a simple wall-mount/tabletop X10 controller that performs simple functions like this, and they're quite inexpensive. You can spend anywhere from about $25.00 to $200 on a simple piece like this, and programming it is easier than setting up your VCR (fortunately!).

Another thing you can do is designate each room to its own House Code (A through P). The simple plug-in modules in each room can then be distinguished from each other by Unit Code (1 through 16). In this event you must also plug in a Base Receiver for each House Code.

The reason you might do this is because X10 controllers usually have the ability to send ON/OFF and DIM/BRIGHT commands to all devices on a specific House Code as well as to specific devices.

As a matter of practicality, this means that you can control all lights in a particular room with a single remote command ("All lights ON/OFF/ETC."), thereby creating lighting "zones."

Different controllers have different capabilities in regard to this. Keychain remotes will usually control only a couple of device addresses, whereas larger remotes may control most if not all of them.

So far there's really no intelligence involved. Of course you can add timer-based control if you wish as I said earlier.

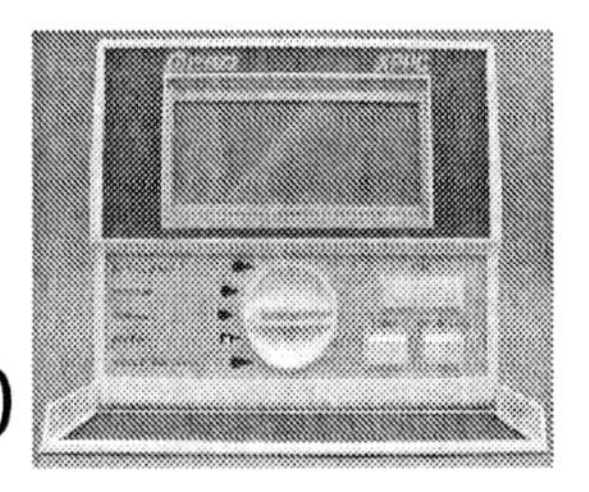

There! Doesn't matter if you're a novice, does it? Anybody can do this! Of course X10 has more advanced tools and applications, but once you get the basics under your belt it simply becomes a matter of putting things together. And when you add some sort of smart controller (like the JDS Stargate), the possibilities are virtually endless!

A somewhat scaled down version of the Stargate is the JDS Timecommander and Timecommander Plus. You can read about these at www.jdstechnologies.com.

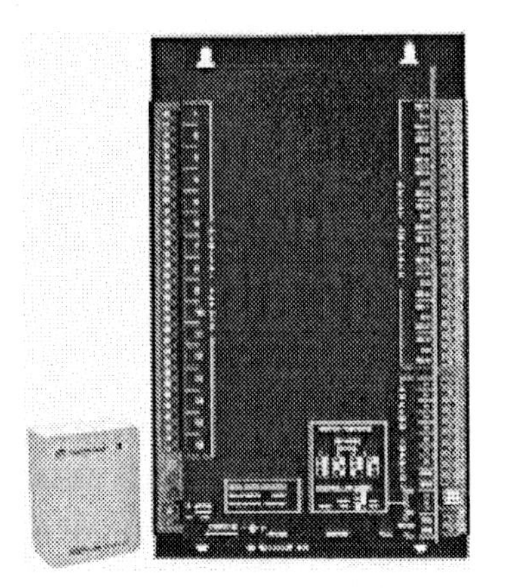

But if you can afford the Stargate, you should go for it. It adds a built-in voice-mail system (and the corresponding control over your home from the telephone (locally and remotely).

JDS has also designed some keypads to work specifically with Stargate. All three can be expanded to include infrared control (very useful!)

CONTACT CLOSURE:

Think about a light switch for a moment. It's basically a loop of electricity flowing from the power company into your house on one wire, through a switch, then a light bulb, and back out to the power company again *(I know - household current is AC and I'm describing DC, but hey, it's just an illustration)*.

When the light switch is turned off, the wire is essentially "broken", and the flow of electricity is interrupted. When it is on, the circuit is restored, and electricity flows through the light bulb again. That's all that Contact Closure is - it's either open (off) or closed (on).

In most cases you won't be using high voltages (if you do, you'll want to enlist the aid of a qualified electrician). Low voltages applications are generally pretty safe, though.

A good example of low-voltage contact closure can be found in a typical hardwired security system (see the chapter on Security.

The sensors that are installed in windows and doors are nothing more than normally-open (n.o.) or normally-closed (n.c.) contacts, with safe low voltages coursing through their veins.

Even Motion Sensors work like this. The security system can detect whether the contact closure has occurred by the presence or absence of current returning through the loop. If it expects to see a continuous flow of current (n.c.) and somebody opens the door - whoops! The circuit was just broken - and the alarm countdown begins.

One of the neat things that you can do is to integrate the infrastructure of your security system into your automation design. Somewhere in the security sensor loop(s) if you install some sort of I/O (in/out) device, you can use the security system to inform your "brain" of what is going on with doors, windows, motion, etc.

I know I'm getting a little ahead of myself now because I'm touching on intelligent design, but just to give you an idea of what I'm talking about:

One night I wanted to surprise my wife with a gift, so I bought her a dress and rigged up a little "show". When she arrived home from work the security system detected her entrance. Getting it's tip from that, a voice-over played through the stereo speakers (my voice, of course) announcing her little gift. Timing it just right, an overhead light then gradually came on in the kitchen where she could see what was waiting for her.

If you're a guy, this is how you get your wife interested in your hobby. . .

. . . a pretty important point if you really get into this stuff

OTHERS:
DTMF is simply a reference to touch-tone telephones - you know, the different beeps that the phone makes when you press numbers. You can assign values to them for a number of smart controllers so that you can use your telephone to issue commands.

RS232 is a protocol generally used by the serial port on your computer.

RS485 is another protocol used over some hard-wired systems. Again, it's not critical to know everything about all of these. It's only important that you don't find yourself too intimidated to use them.

Anyway, that's enough background info for now. Let's get into learning some things. . .

The Major Subsystems

There are a lot of different directions that you can go with this. For practicality's sake, I'm going to discuss these in general terms, and *then* we'll put them all together using specific items.

Remember the quick little X10 project we did earlier

(turning the light on from the car)? That was about as simple as it can get - in fact, I don't know if I would call it "automated". It was simply remote control of a light (but in some cases that may be all you need).

Here is the "modular" approach I spoke of earlier:

We're going to look at several different kinds of "electronic systems" and then - when we're all done with them - we'll talk about INTEGRATING them! If you should choose NOT to go to the expense of integration (e.g. via Stargate), you can still make use of these now.

You'll notice that I have more to say about some of these than others. Here's a list of some of the different systems we'll look at:

- X10 (using existing household electrical wiring)
- UHF
- Infrared system (usually for distributed audio/video)
- Distributed Audio/Video (for your infrared, of course!)
- HVAC (heating and air conditioning)
- Security system

Subsystem 1: X10

X10 signals are passed in a couple of ways. We already saw that you can use a UHF remote to manually transmit an X10 signal to a receiver module - But if you're using a wall-mounted or table-top controller that's plugged in, you'll be passing X10 signals to the module(s) through the electric wiring.

One thing you'll want to be careful of is what you plug in to your X10 modules. The typical X10 module/switch is designed to control incandescent lighting. Any X10 module which has dimming capabilities can do ***damage*** *to electronic devices which don't dim.*

What doesn't dim? Most fluorescent lights, or anything else that's not a light! If you want to use X10 to control something else, use an APPLIANCE module, which has only ON/OFF capabilities.

Actually, there is an "inductive" X10 device which can control ceiling fans, etc. (things that don't "dim", but have variable motors).

Make sure that you use the correct device for your needs.

If you're passing signal over house wiring you may encounter some bugs, especially if you have a larger home. Occasionally you might find that the signal seems to get lost. There are usually reasons for this that we can work around, though.

I'm not an electrician - however, I've been forced to learn a few things along the way which have helped me (as they will you) in troubleshooting.

First, you should know that household AC current has two "legs" or "phases". The X10 signal has to pass from one leg to the other in order to complete its path, and unless you have a 220V appliance turned on (like an oven or dryer) which "bridges" the two legs, the signal has a long path to travel. It actually has to go outside to the power company's transformer and back in to your house. If the distance is too great, the signal may die before it gets back to your house.

A good solution for this is to have a **"signal bridge"** installed at the breaker box in your house. That way you limit the distance the X10 has to travel.

area. A pre-assembled 4X6' backboard should be installed to secure the prewired RG6 Quad Coax, Cat 5, security and smoke detector cabling to their appropriate positions in preparation for the installation of the Security, Cable Distribution, Phone, PC Network, and X10 Power line Carrier (PLC) Systems, when the residence is completed.

- Twin RG6 Quad Shield coaxial cable should be "home run" from each applicable room back to the HCC for distribution of cable, Satellite, DVD programming and entry video cameras.

- Twin Category 5 cable should be "home run" from each room planned with a phone, PC, system wall controller and thermostat, to the HCC. The Ethernet LAN will be established on the Cat 5 and the Ethernet Hub will be located at the HCC.

- The Security System door / window contacts, glass break detectors, motion detectors and smoke detector wiring will be home run to the HCC.

- High quality jacketed 14 and 16 gauge speaker wire, such as ProFlex & Monster should be prewired to each of the rooms and exterior areas of the home identified with speakers, from the area identified for the home entertainment center.

- A Cat 5 IR system control cable should also be prewired from the planned A/V equipment rack area to those areas selected for wall controllers of the whole-house audio system.

- An RS232 cable will be prewired from the primary PC to the Stargate area on the HCC to permit remote programming and control.

- Your HA installer must coordinate with the electrical contractor to insure that the Line, Neutral, & Switch Legs are available at all switch points and that only deep switch boxes are used to insure that any Smart switch & receptacle circuits can be included in future X10 PLC Systems. A minimum of 12 receptacle outlets each shall be installed at the HCC & A/V areas.

- A whole house surge protector and an X10 signaling Coupler / Repeater should be installed by the electrical contractor at the electrical panel to establish an X10 Power Line Carrier signaling network.

Security, Smoke Detection, & Access System
We recommend that a central station monitored Napco Gemini security & smoke detection system be installed. This is a Hardwired / Wireless multi-tasking centrally monitored system. Monitored contacts should be installed at all first floor and basement exterior doors and windows, as well as glass break detectors, interior motion detectors, heat detectors and smoke detectors. Plain English keypad stations should be located at three locations including the Master BdRm.

The Security system should be integrated into the JDS Stargate system at the HCC for automatic control of safety lighting when the security system is tripped.

A 300' range Street Smart ultra-secure coded remote control system with (2) 4 Button Key Chain remotes can be provided for control of gates, garage doors, interior doors, interior and exterior entrance lighting and the security system from arriving vehicles. The system can be expanded to include exterior motion detectors or driveway vehicle detector probes integrated with Stargate at any time.

Distribution System for Cable, Satellite & Camera Video + PC Internet Network.
A Channel Vision RG6 Coax Cable & Category 5 Cable distribution system will be installed at the HCC. This Panel combined with Xantech Infrared components will perform the signal amplification, modulation and distribution of the video signals from of Satellite, Cable programming, and entrance cameras to those rooms planned with a TV or PC. An IR target will be installed in the rooms planned with a TV to enable video source selection and control of Satellite programming, Cable & the Entrance Camera. If a Dual LNB Satellite Dish is utilized this will allow connection of 8 or more Satellite Receivers with individual control. Programmed Universal Remotes such as the SL9000 or PUR08 should be used for control of TV, A/V, and lighting.

The Cat 5 distribution portion of the panel provides the convenience of a patch panel to distribute voice and data to each room directly from the phone system. A PC LAN 8 node Hub will be included in the panel. High Speed Internet Cable Service or DSL Service can be established on the network to allow all family members to be online simultaneously. Multimedia outlets installed in each room will provide phone service, video, and PC Network connection data where required.

Phone System
A Panasonic Advanced Hybrid Phone System should be installed and integrated at the HCC. The system is expandable to 6 incoming lines and 24 extensions. The system will include Caller ID, a door phone at the front entry, hands free intercom & paging. The available phone handsets include 90OMHZ cordless system phones, corded LCD and non LCD system phones.

The 900MHZ cordless phones offer the convenience of remote and local control of anything electrical from any point in the home or on the grounds.

Great Wireless Phone System alternatives to the Hardwired Panasonic Phone System would be the two incoming line, 10 extension CyberGenie with voice recognition & PC/Internet integration, as well as the two line, 8 extension Siemens 2.4 GHZ Gigaset which is extremely clear, high performance, and attractively priced.

Home Control and Automation System
An X10 PLC network will be created on the house electrical service using Surge Protection, Signal Couplers, Amplifiers, and noise filters as required to ensure rock-solid system operation.

A JDS Stargate, the most comprehensive and best supported integrated home controller available will be installed. Stargate is a Windows based, PC programmable home control system that interconnects the Phones, HVAC, PC,TV, Video, A/V, Lighting and all other home systems, allowing them to communicate with one another for control purposes. In addition to X10 control offered by Stargate, it also includes digital and analog inputs as well as relay contact closures for reliable interfacing to all other home systems. The Power Line Carrier (XIO PLC) network will be established on the electrical service to enable control signaling on the AC power line throughout the residence and exterior lighting and recreation areas. Phone control is integrated into Stargate turning any touch tone phone into a great system controller. Voice response and Caller ID are also integrated in Stargate. Check Stargate out at www. jdstechnologies.com.

Smart Wall Switches / Modules will be located to insure adequate safety, security, and scene lighting. TouchScreens, and 4 Button Leviton Wall Controllers will be installed to enable automatic and remote control of selected interior and exterior landscape and pool lighting, house audio, and status conditions. This insures that the residence will always present a lived in appearance. The 4 Button Wall Controllers and TouchScreens also provide a means to trigger Scenarios such as Party Time, Away, Movie Time, Good Night, I'm Home etc. that involve several home systems automatically with one button control. Voice Response can be provided for Scenario selections, Guest Arrivals, and Wake-ups etc as desired. An RS232 cable should be routed from the Stargate at the HCC to your primary PC for both control and programming from your PC.

Optional Home Control Sub-Systems:

- Dimming Smart Switches can be added to create "one touch" scene lighting, as can additional TouchScreens and Wall Controllers.

-An Irrigation Controller, controlled by Stargate can provide convenient integrated lawn & garden automatic irrigation.

-Smart bi-directional Thermostats can be installed to offer centralized control based on time of day selections, or remote & local phones and or universal hand held remotes.

-Voice recognition software can be installed on the primary PC and can be integrated with the Smart Home phone system to easily trigger scenarios such as Good Night , I'm home etc.

-The system can be expanded with an Infrared Expander to allow "one touch" control of whole house audio systems and home theaters with the same 4 Button Wall Controllers and TouchScreens.

- A Pool/Spa Controller can be installed and integrated with Stargate to offer complete system control.

Whole House Audio System

A Niles or equivalent based audio system will be installed at the designated Entertainment Center location. The system can will consist of a AVR Audio/Video/Receiver, Niles Flush Mounted In-Wall and In Ceiling multipurpose speakers and Niles OS10 exterior speakers with Niles RVL6 speaker zone, source selection, and volume controls. This system can be upgraded to include other amplifiers or speaker types as well as by adding room TouchScreens or KeyPads for remote source selection and volume control in place of the proposed room volume controls.

This system shall be integrated at the HCC to permit Stargate Voice Response through the house audio system for audio caller ID, audio reminders, and house status such as the temperature in each room security status, garage door position and more as desired.

There was actually more in that article than needed to be covered at this point, but it's good reference material.

We've looked on some pretty simple applications for X10. If you're looking for something a little nicer, you might consider X10 wall switches and/or outlets. This way you don't have to see the clunky little modules everywhere (for plug-in lamps or appliances, you need either the outlets or modules. For lights that are switched, you obviously need switches).

Again, you can spend a little or a lot. You can buy some fairly cheap button-type wall switches from X10. Beyond that, Leviton makes some moderately priced paddle-style wall switches, but to me they feel a little mushy. You have to press the bottom once to turn the switch on, and press it again in the same place to turn it off. If you press and hold, the light will brighten or dim, depending on the last thing you did.

Until Grandma gets used to it, she'll be a little confused.

I'm not necessarily knocking the Leviton switches, because I've used them a lot. They're affordable.

But if you would like something still nicer, you'll really like the PCS brand switches. They're on the expensive side; but the nice thing about them is the way they work *manually* (like ***normal*** switches). They are true "rocker-style" paddle switches (press top for "on", bottom for "off"), and they have a degree of intelligence built into them. You can easily program them for a preset brightness level for every time you turn them on - you can even program them to communicate with each other so that you can control an entire bank from just one switch (up to 16 different lighting scenes).

If you really get excited about your lighting system, PCS has an "LCP16-8000" lighting panel. This steps you up to the benefits of a hard-wired lighting control system, complete with "star" wiring topology. It does, however, still interface with Stargate via X10.

One other somewhat pricey alternative is Lutron's

Homeworks lighting system. It also utilizes the "star" wiring for high-voltage electric, but then connects from the central panel to switches via low-voltage wiring. And rather than interfacing with Stargate via X10, a single Cat5 wire connects Stargate with the Lutron panel (RS232 connection).

What's cool about both the PCS *AND* Lutron system is that Stargate has some specific features built into its software just for Lutron and/or PCS. They're tailor-made for each other!

... Anyway, getting back to some things a little more basic...

I used to be a fan of Switchlinc switches, and I guess to a degree I still am. They're very similar to PCS and they come in a "lite" version which can save you money. However, occasionally bugs seem to pop up on some installations. At this point it's rumored *(yes, on the internet)* that Smartlinc (a.k.a. Switchlinc) denies the problem, but it *would* explain some odd things that I've seen.

Be careful you don't make the mistake that I did. Make sure you count the total wattage on the lights being controlled by your switch. The average switch can handle a load of about 600 watts (check its documentation). If you have a bigger load than that, make sure you buy a higher grade switch. On an extensive installation for one of my customers, I must have burnt up a half-dozen switches before I realized what I had done.

I told *you I'm not an electrician, didn't I?*

One thing you should be aware of: If you get an electrician to install these switches, make sure he studies the instructions. Some of them don't "wire up" like ordinary switches, and many of them also *require* a neutral wire at the switch box.

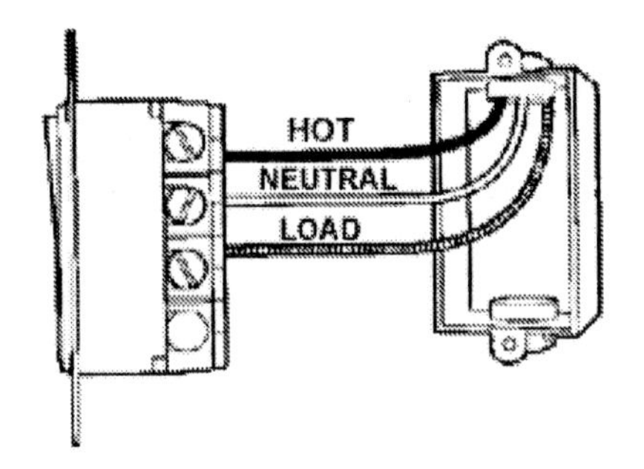

You should also be aware of some terminology. A **three-way** switch typically refers to a light that is controlled by two switches (three means two - get it? Well, I don't ... but that's how it is).

But with smart switches, the terminology is usually "master" and "slave" switches. For any given light (with two or more switches), there will be only ONE master. The others will be the slave switches. Fortunately, the slaves are not so expensive.

Subsystem 2: UHF

I said earlier that there isn't much use for UHF, but that's not exactly true. There are a million different applications, but usually you'll find yourself making a number of miscellaneous additions to your system, rather than what you might call a "single UHF subsystem."

Maybe that's just a way of looking at it. But there are wireless ways of doing a lot of things: communication between your RF X10 remote and base receiver is obviously wireless; Driveway vehicle sensors, door cameras, security, audio & video ... you can even do a wireless LAN for your PC's.

The difference is: **wireless is easier** (not always cheaper, but sometimes the only practical way to do certain things), but **hardwired is usually more reliable.** In a while, we'll make use of some of these applications.

Subsystem 3: IR

I've already described IR to UHF to IR (remember the RCA Powermid Cones?). You can use this if you wish, but I'm going to talk about a hard-wired system.

In the custom A/V industry, there are a several known systems, among them Niles, Jamo, and Xantech. They all are built pretty much the same, though you generally ***cannot*** mix and match their parts.

Let's use a stereo system as an example, with speakers scattered throughout the house (since that's the most common application anyway). You wish to have control over the different components in remote rooms, and it basically works like this:

Located back at the stereo, there is an IR **"switchbox,"** or "connecting block." From this device, tiny IR **emitters** (or flashers as they're sometimes called) lead out to the various components. Each one either sticks directly to the front of the component's IR window, or it may take the form of a larger "flooding" flasher which can hit several components at once.

Then, wire runs go from the switchbox to your remote locations where you'll need control. These runs can terminate in table-top **IR receivers**, or as I prefer, in-wall IR receivers which fit inside a standard J-box.

Manufacturers typically recommend that you run shielded 1 pair wiring with a ground (3 wires). The shield is mainly to protect you from EMI (electrical) interference. However, I've found that if you're careful in how you run your wiring, Cat 3 or 5 works just fine. Just be sure that you don't make your runs directly next to electric (especially in parallel with it!)

The other potential problem you might have comes from sunlight. Don't locate your sensors in direct sunlight (or even in brightly reflected sunlight). It can drive your IR system crazy!

Once you power up the switchbox/connecting block, all you have to do is aim your remote at the IR receiver. Commands are simply relayed back to the switchbox, which in turn passes them on to your stereo.

This is fairly simple, right? If you wish, you can replace the IR receiver with a Niles Intellipad or Xantech keypad. Among others, these two are simply hard-wired learning remotes. Follow instructions for programming them, and *you're done!*

Later on, we're going to add another twist when we integrate this concept into our whole design. Here I'll not be using either the Niles or Xantech keypads (so don't rush out and buy them yet). Instead I'll show you the JDS LCD Keypad and IR-Xpander to introduce intelligence into the system.

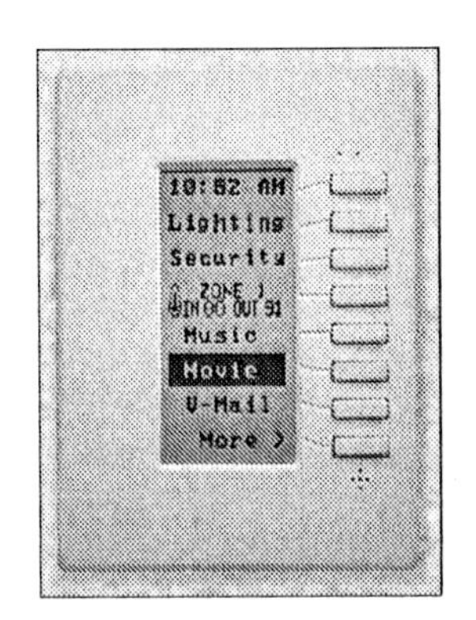

Wall-Mounted LCD Keypad Interface for Stargate

Infrared Xpander

Subsystem 3: Distributed Audio/Video

Subsystem 3-1: Distributed Audio

Let's start with a whole-house audio scenario. First of all, most stereo receivers can power no more that one or two sets of speakers (usually marked "A" and "B"). So if you have multiple rooms which you are going to connect, you need an "impedance matching" device (also known as a **Speaker Selector** box) so that you don't overdrive your amp.

You want to make sure that you get a quality piece here. Without naming well-known names, just remember that if it's cheap, **it's *cheap!***

To begin your wiring, you need to choose your "home run" location (where your speaker selector box will be). If possible, make this the same as your stereo location. You'll need to pull either a 4-conductor speaker wire or two 2-conductor wires to each room. Do not try to wire these in series unless your equipment specifically tells you to do so.

A really elegant approach is this IR-controlled speaker selector box (see picture below). It controls up to six pairs of speakers, & you can remotely control which speakers play. It's a little more expensive, but will give you some "zoning" control.

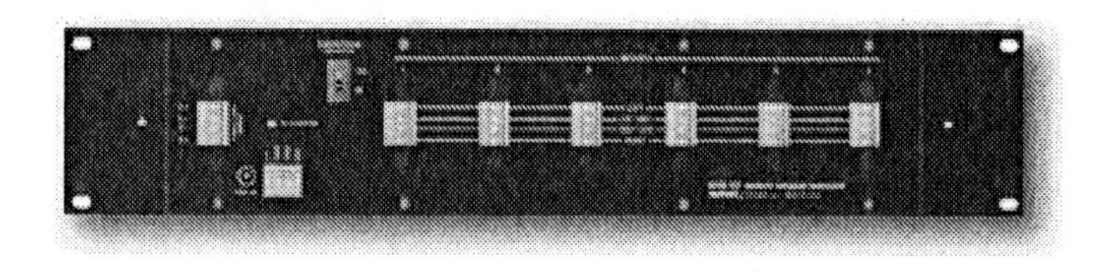

I would recommend at a minimum 16 gauge speaker wire, but 14 is better. And if you can get wire that is paired and twisted, you'll be ahead when it comes to avoiding interference. Also, check the strand count if you shop and compare. Generally, the higher the strand count, the better the wire.

In order to control volume independently in each room you need volume controls (and again I'd not recommend that you go really cheap). It's possible to have these located right on the speaker selector box, but this obviously robs you of a lot of control. It's better to have them in each room, either as a table-top or in-wall version. If you use the in-wall type, you could simply install a double-gang box which will accommodate both an IR receiver and volume control. Personally, I like to locate these near the light switch as I enter the room, but whatever you like is fine.

The wire coming from the speaker selector box terminates in the volume control. You then run 2-

conductor wire to each of your two speakers in the room.

You can also install "impedance-matching" volume controls. They cost a little more, but they do away with the need for speaker selector boxes.

By the way, the same rule applies with speaker wire and electric as it did with IR wire. Keep as much distance as is practical, especially in parallel runs. Otherwise, you *could* get an annoying hum in your speakers.

One other point: Low voltage lights and ceiling fans have been known to introduce hum as well. So if you're doing new construction, you might want to ask your electrician to put your stereo system on its own circuit to avoid this little problem.

ABOUT IN-WALL SPEAKER PLACEMENT:
It's generally best to have a pair of speakers facing the same direction, especially if they play in stereo. The effect is not all that pleasing if they face each other, or if they're at right angles to each other.

Because in-wall speakers can be situated in the ceiling as well as the wall, you should consider the effect you're looking for.

*If you want a nice background feel for music which isn't necessarily coming from a particular direction, put them in the ceiling (they're also less conspicuous this way, which makes interior designers **AND** wives happy).*

On the other hand, if you'll be doing intense listening which requires a good reproduction of the stereo soundstage, put them in the wall. If you like your music and aren't too concerned about aesthetics, do it this way.

Distance between speakers matters, too. A good rule of thumb is to imagine a triangle with equal length sides. The listener sits at one point of the triangle, and each of the speakers sits at the others.

In other words, there should be approximately the same distance from your ears to each speaker as there is between the speakers.

If they're too far apart, your "stereo soundstage" will be very tiny. If they're too close together, you'll hardly have one at all!

One other option you have if you're majorly concerned about aesthetics: Soundadvance (and a few others) makes speakers that completely disappear into your drywall. After they're installed you can spackle and paint over them. I installed a few sets of these several years ago (though I don't think they were from Soundadvance), & they sounded really good. But from a "service" point of view, I can imagine potential nightmares. If you install these, you better make darn sure you remember exactly where you put them!

Whew! That was too much gray type! Umm - where was I?

Oh, yes. Once your wiring is in place, it's just a simple matter of making all the connections. Don't be too intimidated by the back of your stereo/ surround receiver *(the real learning curve is probably with the crazy remote - something we hope to simplify later).* If you have a plain stereo receiver, you have your "speaker-level" connections (typically marked "A" and "B"), and your "line-level" connections for your different components.

If you have a surround receiver (for Home Theater applications), it won't be much different except for the speaker connections. You'll probably see speaker-level connections for rear and center channels as well as A & B (for the front speakers).

There are so many variations on receiver & amplifier capabilities that there's no way I can discuss them in depth. Chances are you don't need much input from me on that anyway since you're the one with the manual (I hope!).

Subsystem 3-2: Distributed Video

When we talk about distributed video, it could be in regard to a number of different sources. We may be referencing the distribution of *Satellite TV, VCR, DVD, Cable* & the like ...

Or maybe your concern will be how to view security cameras on different TV's. In either case, the general principles are the same. So ...

It's time to get acquainted with a device known as a ***modulator.***

A *modulator* takes a video source from one frequency (or *channel*) and converts it to another.

Think about your common VCR. What channel do you put your television on to watch a tape on the VCR? Probably 3 or 4 (you even get a choice!).
And if you've ever used the VCR's tuner (changed channels on the VCR) to watch cable or off-air stations, you STILL leave your TV on channel 3 (or 4). So what's going on ...?

What's happening is that the VCR is ***modulating*** outgoing video to channel 3.

So if you have a number of video sources you wish to distribute throughout the house, a modulator will help you accomplish it - even if by default they all are generated on the same channel. Let's use security cameras as an example:

And again, there are a variety of cameras available, but let's be generic for ease of illustration. In my home I've run coaxial cable plus category 5 for power (a bit of overkill with that cat 5) to each camera location. When I mount the cameras on the outside of my home & make all the connections there, down in the basement my coax terminates into a modulator. For each camera that enters the modulator I choose a unique *unused* channel for that camera to be distributed on. The modulator you choose will give you precise directions on how to make those settings.

Now you have to **inject** the video signal into the coax running throughout the house. Here's how ...

If you already have coax cable for any video source being distributed to different TV's, you will notice a "splitter" (we'll call this splitter "A") where all the cables meet. Basically the splitter takes your cable source and splits it (*so **that's** why they call it a splitter!*).

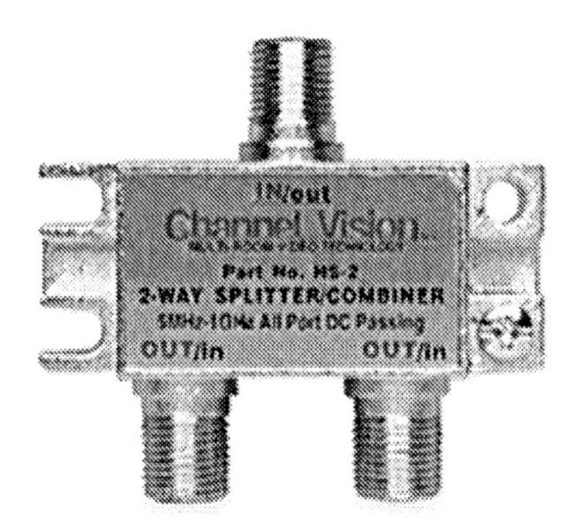

What you want to do is a sort of reverse process of *combining* channels. Take your coax sources after they exit the modulator and plug them into the "output" of a splitter (you may have to make sure your splitter will accept signal flowing in the opposite direction).

Continuing on, you come out of the "input" of the splitter back into the "output" of another two-way splitter. Your previous cable/ antenna/satellite source (before it gets to splitter "A") goes into the other "output" of that same splitter.

(Man. Hope this isn't too confusing ... just one more step:)

Now - from the "input" of that two-way splitter you want to go into your (already existing) splitter (splitter "A") which distributes video to all your televisions. And ...

you've got modulated video throughout the house!

---

*Later on, we'll discuss how to make access to those security cameras **automatic** through integration of motion sensors, doorbells, etc. At one of my customers homes, he wanted to be able to tell whether someone was approaching the house when he was watching a movie in his media room downstairs. After all, with his THX system blasting he was pretty much oblivious to what was going on in the next room, much less his front yard.*

When all was said and done, movement at the front door paused his movie and switched the projector to a view of the outside camera. Other cameras responded differently to different circumstances, depending on darkness, whether the homeowner was actually home, time of day, etc. Pretty cool!

Subsystem 3-3: Home Theater

Yep, this is the big daddy. Lots of companies make their living doing ***only*** this. The reason is that this is where your big money could *potentially* be - but *that* depends on your budget and your wants/needs.

You don't necessarily have to spend a fortune. Decent sound from a *surround receiver* and good speakers can make a modest television room seem pretty grand.

Speaking of Video:

However, if you're into some more elaborate setups *(I'm not gonna go crazy here!)*, you could find yourself considering some alternatives. A **rear-projection** television isn't a big step up, but it gives you the potential for a larger screen. However, a **front-projection** television puts you into the category of - oh - *maybe* big time, depending on all things considered.

In case you're wondering, I'm talking about a projector

that mounts to the ceiling (or sits on a table) and "projects" onto a screen. This is where you have the potential for 100-inch + screen sizes ... and **BIG** bucks.

But if you can afford to do it this way, you'll probably want a dedicated room, or at least one that can be darkened. The reason for this is that you won't get nearly the brightness on your screen that you'll get from a conventional tube TV.

Also, there are some choices between the kinds of projectors available. It used to be that you had to choose between a **CRT** (or, "3 gun") projector and a simple **LCD** projector. The CRT produced a better quality picture, but it was difficult to set up and maintain. The LCD projector was very simple. You aimed, turned the lens to focus, and that was it.

... But the pixels were obnoxiously evident.

Now the market has produced the **DLP (Digital Light Projector)**. It has all the advantages of the LCD in terms of simplicity, and ***almost*** the quality of the CRT's. Unless you're a real purist, I'd recommend going this route. Just expect to pay a minimum of $3,000 for it, maybe more.

And that's just for the monitor. You still need an audio source (surround receiver) and video sources/tuner (VCR, DVD, etc.).

So, ...speaking of audio - there are several sound formats available to the home theater enthusiast. The old ***Dolby Pro-logic*** is still around and popular. It separates the soundtrack into 4 channels and also sometimes has a subwoofer channel (hence its name "4.1"). The channels are:

- Right front
- Left front
- Center
- Rear

The newer and better sounding format is ***Dolby Digital (AC-3)***. This is primarily for use with the newer DVD's or other digital formats. Its main advantages are that it features better separation between the channels (a more distinct sound), and the rear channels are "stereo" (instead of Pro Logic's single channel for both rear speakers). Therefore, you'll sometimes find it referred to as "5.1."

There's even a "7.1" system available if you want, though when you get down to reading the fine print, two of the channels are comprised of a matrix in many cases (i.e. a single channel electronically separated into two). At the time of this publication, the only TRUE 7.1 system that I'm aware of is from Lexicon.

If you do much serious shopping around, you'll probably come across the term **"THX."** That happens to be a reference to a quality standard set by Lucasfilms.

The THX label isn't a brand - but it *is* an assurance of high quality which may apply to both components and/or speakers. Here's where you may run into "7.1" systems (an *additional* pair of speakers on the side/rear added to the setup).

It's great to have this mark on your equipment *(if you've got the money to burn)*, but don't let anyone convince you that you're necessarily lacking in quality without it.

Some manufacturers don't seek the THX stamp because it's expensive for them to obtain it, and it may unnecessarily drive up the cost of the equipment.

What to look for in a Receiver:

It's totally up to you how much you spend on a receiver, or whether you buy "separates." But you **do** have to be careful when you compare "specs."

Watch it when a receiver is advertised as having a gazillion watts of power. That often doesn't mean as much as you might think. Does that number reflect ***per channel?*** Or is it the sum total of **all** channels? If they make a big deal of it, it's probably a total.

Also, a lot of power consumed doesn't necessarily equal a lot of power used efficiently. For instance, some of the smaller Harman Kardon receivers (if you ask me) are much to be preferred over certain bigger rivals.

One other thing to be aware of is that some manufacturer specs aren't necessarily comparable. They don't always use the same calculations to arrive at their numbers, so if you aren't careful you could be comparing apples to oranges & not even be aware of it.

If you want to shop for a brand name, be aware that some manufactures make both low-end **and** higher-end stuff. Generally the more commonly marketed brands will do this. Custom lines like Denon, Harman Kardon, etc. generally only make stuff with "good" materials. They tend to be more pricey than Radio Shack stuff - anyway, do as you will. I'm only telling you that you'll probably get what you pay for.

Speakers:

About the only thing I think I need to say here concerns the subwoofer. You can get a **"passive"** or a **"powered"** subwoofer (powered has its own built-in amp & other controls). The powered sub is what I'd recommend for the theater applications, while in my opinion the passive sub is suitable for music.

Not only does a decent powered sub have a volume control, but it usually also has a variable crossover control. This lets you adjust the frequency at which the sub begins amplification so that if, for instance, voices are coming across as too "boomy," the crossover adjustment can compensate.

Subs are available in both free-standing and in-wall versions. If you want my opinion, the in-wall subs just don't quite measure up to what you should expect. I'd recommend a free-stander.

Subsystem 4: HVAC

That's "heating, ventilation, & air-conditioning" in case you've always wondered. You can buy standalone multi-zone systems, but again - if you already have a smart controller, you don't *need* to spend all that much.

That's the approach I'll be describing. First of all, you need ***thermostat(s)*** that will respond to your controller. Stargate works well with **RCS** and **Enerzone** Thermostats. In order to keep it simple, I'm going to focus on the RCS brand.

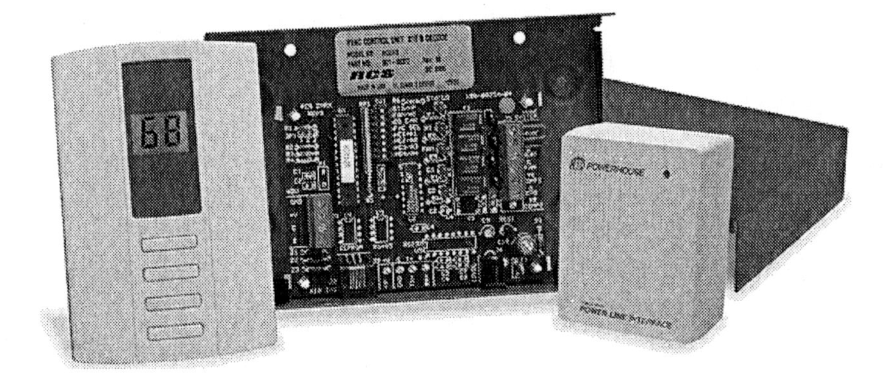

You can get RCS thermostats that are controlled by either **X10 commands** or the (hardwired) **RS485** protocol.

The difference? The RS485 type requires a Cat5

connection, and it's also a little more expensive, but it has its advantages:

You'll want your Stargate to "poll" your thermostat at a regular frequency to obtain info concerning the current temperature. If you're doing frequent polling with the X10 thermostat, it keeps the powerline busy (it can *potentially* interfere with other X10 activity).

But you can have almost continual polling with the RS485 type and no resulting problems! Also, while X10 can be pretty reliable, nothing is quite as good as a direct physical connection.

The other thing that you can do is to install some ***electronic dampers*** in your heating ducts.

This will allow you to control airflow to certain rooms, thereby saving on heating & cooling costs. You simply have to run a 2 conductor wire from the Stargate location to each damper. When power is applied to the wire, the dampers close *(or open, depending on which kind you buy)*.

So as Stargate monitors the situation, you can direct it to shut off (or open) certain ducts based on temperature, time of day, occupancy, or other events.

You do want to be careful, though, not to stress your furnace by shutting off too much airflow for too long. If that happens to be a concern, you can also install a *"Barometric Bypass"* which will help protect your HVAC system.

---

*There **IS** an X10 trick if you happen to find yourself in a situation where you need to apply low-voltage but can't run the wire all the way back to your controller. With a **Universal X10 Module** and a 24V transformer, you can apply power to your device via X10 commands.*

All you need to do is wire the transformer and Universal Module in series, and finish up at the device in question. In response to X10 signals, the plug-in Universal Module opens or closes its contacts, thereby applying (or removing) power provided by the transformer.

You can use this for control of garage door openers, electronic dampers, or just about any other low-voltage device. Of course, if power is already present on the wire (e.g. garage door opener), you wouldn't need the transformer.

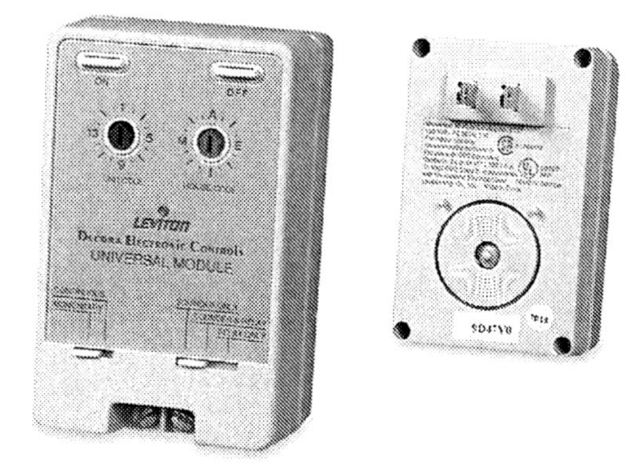

Subsystem 5: Security

A basic security system isn't at all difficult to install. The MOST difficult part is getting the wire where you want it for window and door sensors. As was said earlier, a security system works by monitoring the status of open or closed circuits, and responding accordingly.

If you want a basic security system that works well with the automation features of Stargate, I'd recommend Caddx. In fact, the newest software version of Stargate has some specific code written to work with the NX series of Caddx panels (though you don't have to use that *particular* line).

And again, let me repeat that you don't need or even necessarily ***want*** built-in automation features in your security system. Let the security system do security, and let Stargate do automation.

Below are some pics of the two main types of door/window sensors that you might use. One is a surface mount and extremely easy to install, and the other is recessed (both the sensor and the magnet). The recessed kind is obviously more attractive *(because you don't see it)*, but can be difficult to do in a retrofit environment (after construction).

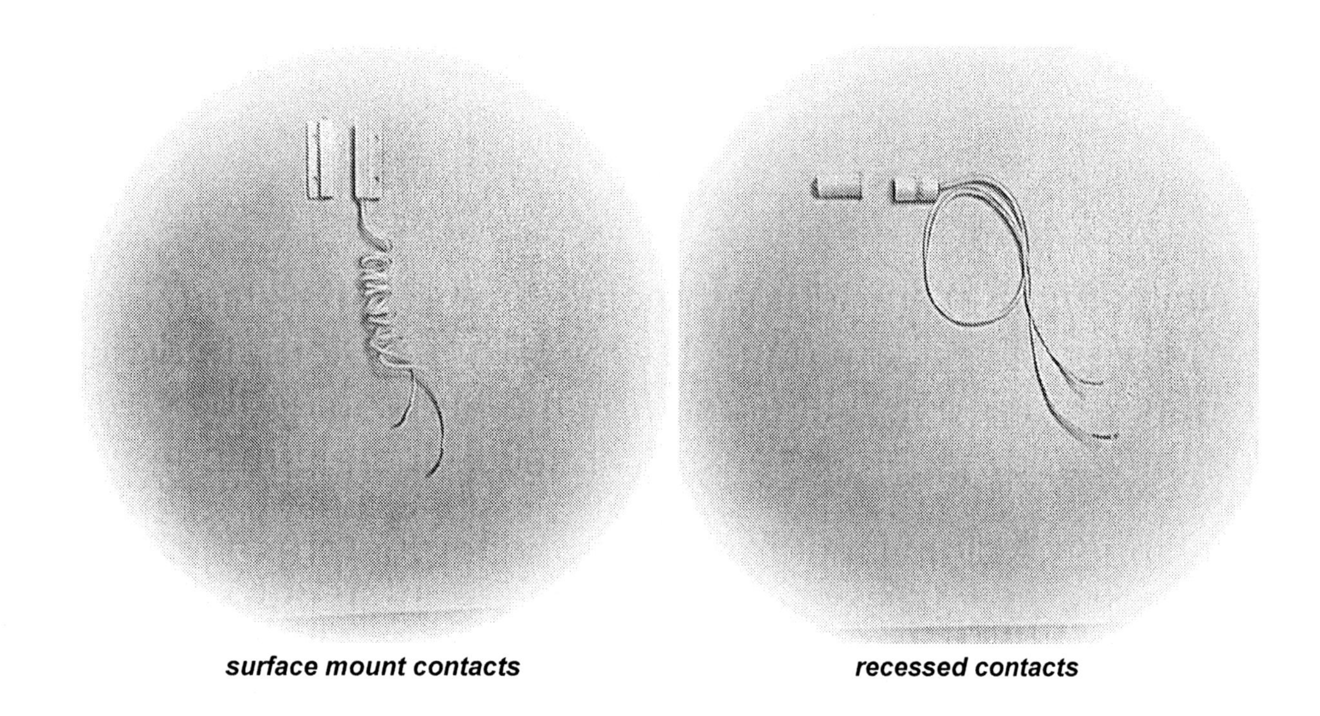

surface mount contacts ***recessed contacts***

It's really important that you grasp how the security panel detects interruption in current flow (if you haven't read the above-mentioned reference to contact closure you should do so now).

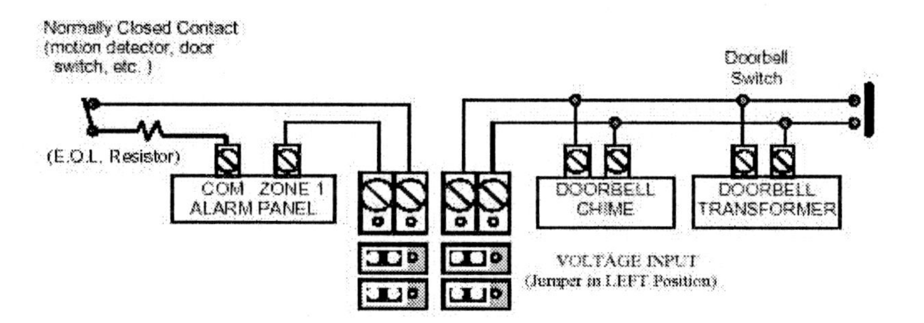

(from Stargate manual: illus. of I/O interface with alarm panel and doorbell)

Notice on the left side of the diagram how the alarm loop runs from the alarm panel *(zone 1)* to the Stargate digital input & back out. It then carries the current to the door/window contact sensor & back to the alarm panel.

All that the digital input does is monitor the presence or absence of current (like the alarm panel). If the circuit is broken by opening the door or window, the alarm panel will react (or not) depending on whether it's armed. The Stargate will also react according to how it's been programmed.

As an extra *(since the diagram shows it)*, you can also see how the doorbell is wired into another digital input. In this case, when the doorbell button is pressed, the circuit is closed and current flows. Stargate senses this, and does whatever you've told it to do.

And in case you're wondering, below is a picture of how to splice the security sensor to your wiring. Since your splice will probably be inaccessible once it's hidden behind drywall, you want it to be a **good** splice *(meaning I don't recommend tape)*. Use "Scotchlocks" or something like it that's designed for permanency.

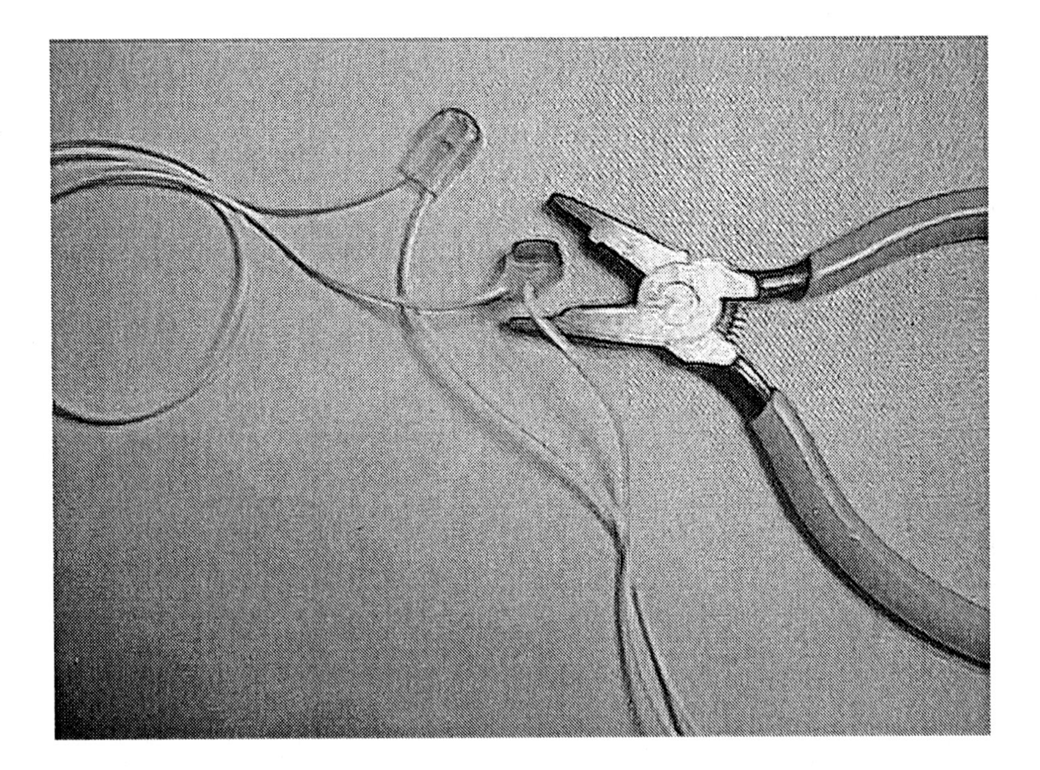

Next is an example of programming Stargate's digital inputs (DI) to take advantage of the security systems sensors. The "trash night" event is utilizing a motion detector (PIR), and the "alarm" event detects output from the security panel when the alarm is activated:

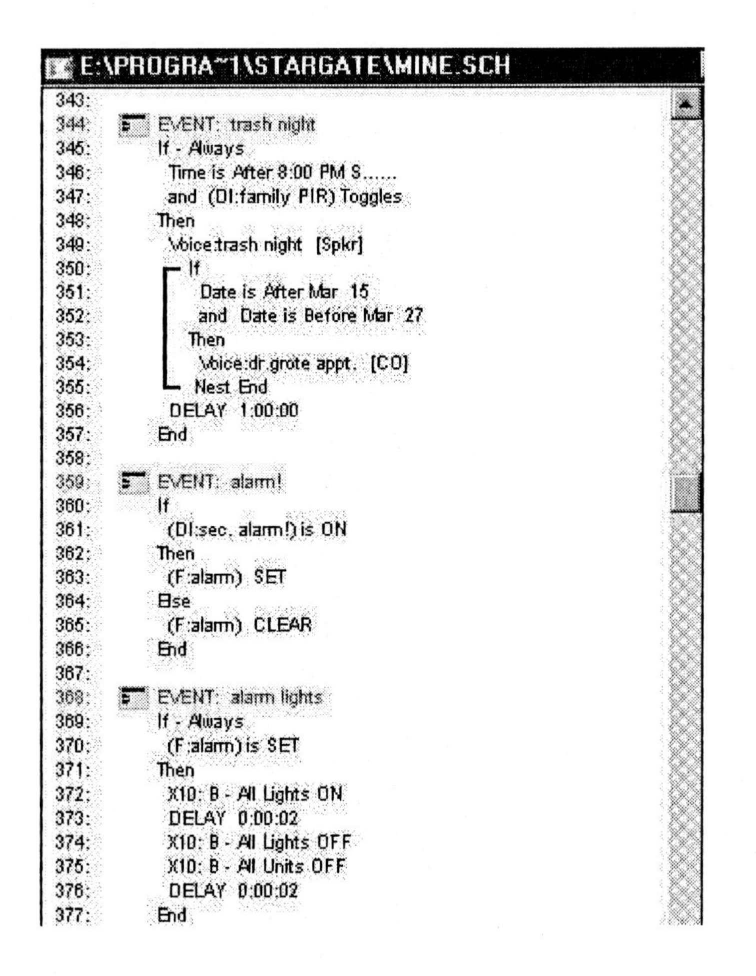

There will be more concerning scripts later. I just want you to understand the simplicity of how security works. Caddx provides easy to read programming instructions for their panels, so you shouldn't have a lot of trouble figuring it out. They're also pretty well known for their decent technical support.

For your consideration, video is increasingly playing an important part in security. While I spoke earlier on distributed video, you might want to consider ***Webcams*** as a vital part of your security system as well.

What's neat about them is the fact that you can upload video clips or snapshots to a remote site when motion is detected. That way, the old problem of securing your video from would-be burglars is averted (they *could* steal your VCR!). If you check out the ***Iwatch*** products, you might find that it's what you need for this purpose.

One thing that's different about the **color webcams** that come in this package is that they run through Cat5 data cabling instead of coax (there's no separate power supply wiring to run). Also, an included card which installs in your PC lets you run the video to your house coax system, where you can inject it into your video distribution. Live audio and built-in motion detection completes the setup!

Part Two:

JDS Stargate

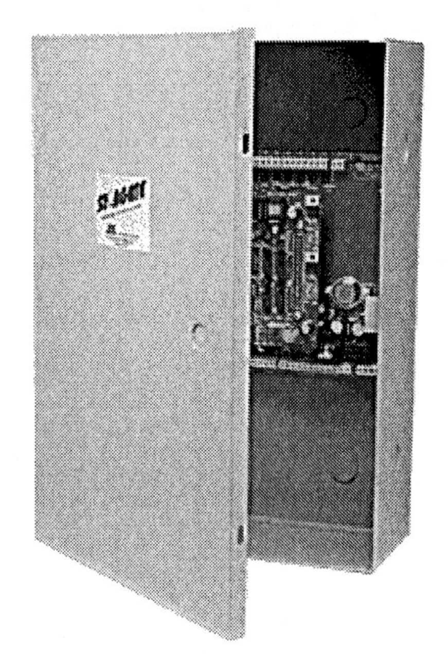

At last we get to the heart of the matter (or really the "brain" of the matter, for that matter). So far we've covered a basic understanding of the major parts of a typical home system. It hasn't been too difficult, has it?

If you can just *begin* to get this next part, you are **well** on your way to a fully automated home. I say "begin" because you don't need to be expert in all the in's and out's of Stargate to make a lot of cool stuff happen. I've worked with (and owned) some devices that were so complicated I found myself really aggravated ...

but this piece of equipment (for the most part) is really simple to program.

However, you'll have a hard time exhausting its possibilities. As your skill in writing scripts grows, you'll find new possibilities that you hadn't considered before.

I don't mean to make it sound like a simple device. There *are* some pretty deep features you can utilize if you happen to be mathematically inclined. While I think I'm pretty good at accomplishing just about whatever I want with Stargate, I have to confess that there are still a few things for me to learn (*yes, even **me**!*).

The Logic of Programming:

Here's how easy it is:

IF it's dark outside
THEN porch light on

It ***can*** be that easy. But what if you don't *always* want the lights on when it's dark? What if you only want lights to come on an hour after dark and go off at sunrise?

It might look something like this:

IF it's dark outside

THEN Delay 1 hour
        porch light on

ELSE porch light off

"IF, THEN," and "ELSE" are the three major commands in writing scripts for Stargate. If you can understand this hyper-technical convoluted geek terminology, you can program!

Jeff Stein at JDS Technologies was kind enough to permit me to post a demo of Stargate's software at ***www.integratorpro.com/stargate/sgdemo.zip.*** That way you can see for yourself some of the features available & how they work.

...Important!

From this point on, it may be a little more difficult to follow if you haven't downloaded the Stargate demo. If you haven't, open your browser now and go to *www.integratorpro.com/stargate/sgdemo.zip.* If by chance it's not there, you can also go straight to *www.jdstechnologies.com* for it.

You'll also need a Zip utility, which you can get *free* at *www.winzip.com.* Be sure to get the free evaluation version if that's what you want.

You may in fact already have a Stargate with the full software installation, and that's great. But what we're going to do first is study the existing schedule and devices in the demo, so it would be helpful for you to have it as well.

So go grab it now ....!

Got the demo? Unzipped and installed? Great! Let's open the program and take a look....

Chances are that when the software opens, you'll need to open the sample schedule and Megacontroller. From *"File"* menu, choose *"open"*, then *"sample.sch."*

Next, go to the *"Utilities"* menu & select *"Megacontroller."*

Finally, choose *"Window"* & then *"Tile"* so that everything is visible.

Very nice. Take a few minutes and explore the different options on the menu and look over the schedule, but don't change anything right now. You can create your own schedule in a bit after you get familiar with things.

One thing you should check out is the *"Help"* menu. Look at *"Help", "Contents"*, and read over things there. There's some good info on X10 theory, but especially note the section on *"Event Basics."*

You'll probably notice that the Help references the TimeCommander instead of Stargate. Don't let this throw you - The software is actually for Stargate, although it is an older version than what you'll be working with in reality (there are a few minor differences which I'll point out).

For instance, here's a screen shot of **version 3.02.** Notice the added features of the "Define" menu compared with what's in the demo:

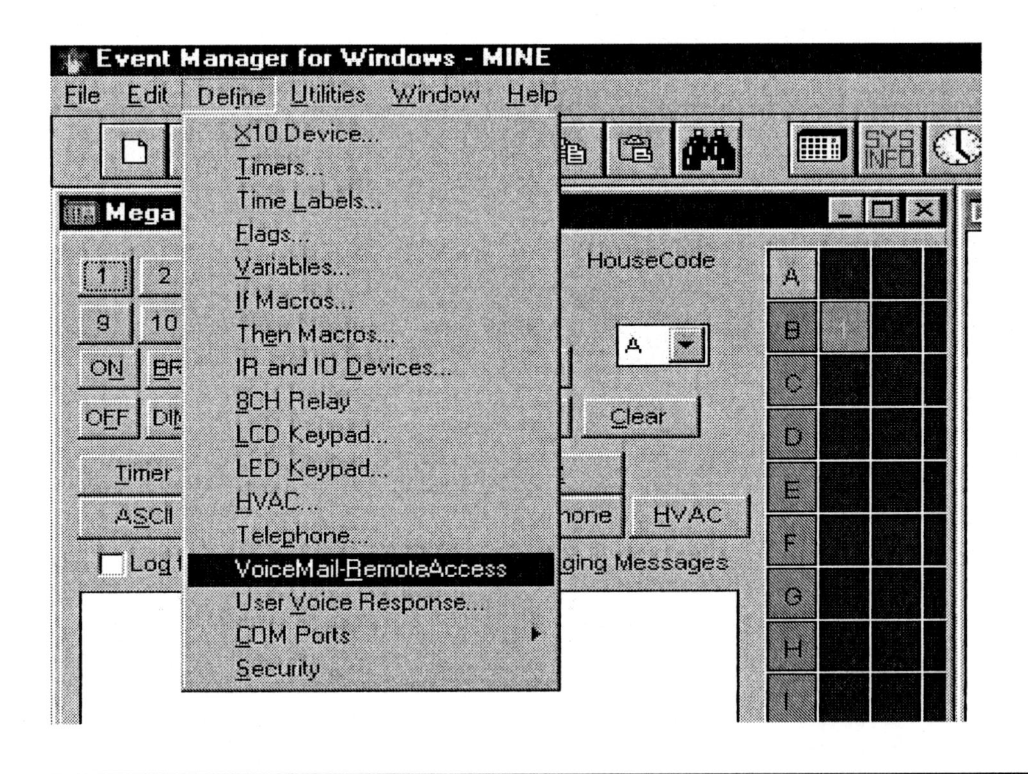

Anyway, at this point **you should take some time to read over this stuff.** There's no sense in my repeating what someone else has already stated "oh-soeloquently."

And if you want, you can go directly to the *JDS website* and view or download a PDF of the latest manual, schematics, application notes, & other stuff.

After you've done so, come back here. I'll be waiting ... and ***assuming*** you've educated yourself on some of the basics....

Later...

Now that you can program - (no, really!)...

Let's begin by taking a brief look at the existing schedule. Please open both the *"sample.sch"* and *"Megacontroller"* so that you can view them side by side (like we did a short while ago). If you've read over the manual, a lot of this should already make sense to you - but let me point out a few things regarding what you're seeing:

You can look closer at any particular line in the schedule by double-clicking on it. This will bring up the actual window where the command was created.

If you'll look at the first **"event"** ***(wake up routine)***, you'll notice what's called a "nested" event. In this case, nested within the THEN sequence is another IF/THEN event. You just have to follow the logic here...

"**IF** the time is 6 a.m. Mon thru Fri, **THEN** (*series of X10 events*) AND **IF** Temperature in zone 1 is less than 75, **THEN** set the temp to 75."

The "nested" (or second) IF/THEN event is only tested if the FIRST set of "IF" conditions occurs (namely, that it's 6 am). You can use nested events like this, up to 3 deep, for some fairly sophisticated logic in your programming.

You'll also want to watch in what order you place your statements. It can make a difference in the results.

Now, look at the second event in the schedule. Double-click on the event title "SIMULATE SUNRISE...."

Among the options available here is the "always" box, which in this case is checked. An "IF-ALWAYS" event is different in that it will continuously test and replay the event over and over as long as the "IF" conditions are true (notice the *delay* statement - it introduces a pause between the repeat of the event).

What normally happens is that when "IF" is true, "THEN" happens only ***once*** (unless the "IF" conditions become at some point no longer true - and then become true once again).

So the effect in this event is that the PCS light switch will *micro-brighten* (something that PCS switches can do) 1 step every ten seconds for as long as the time is between 6:00 and 6:34 a.m. (or until it's at full brightness).

You can use IF/ALWAYS events like this when you have a script that you need continuously executed until acknowledged (like a security alarm?).

Next, if you skip down to line 55 for "EVENT: VACATION MODE SET", we see an instance of a "FLAG" being set...

...This actually brings us to the point of a short discussion of "defining devices."

... Up on top of the screen you'll see a *define* dropdown menu. This is where you select and define the different devices that you'll use in your schedule. In some cases you'll actually define the device in detail (for instance, select *"define"*, then *"Time label"*. Click "new" and you get to name and describe in detail what that time label is.

But for other items, like "timers" and "flags" (like in this case), you merely give it a name. Its actual definition takes place within the schedule itself.

It's really not as confusing as it might sound - just notice that in the VACATION MODE SET event, the schedule tells us ***how*** *the VACATION MODE FLAG is* ***set***.

The following event tells us ***how*** *that particular FLAG is* ***cleared***...

...and the event after that (line 77) is where we finally see defined ***exactly what is supposed to happen*** when the

FLAG is set. Everything (except for the name) is defined within the schedule itself!

The only other thing of note about the VACATION MODE SET event is on line 61. Here, it executes an **ASCII string** (simple text) which triggers a ***.wav*** (sound) file from the PC (instructions on how to do this are in the help menu).

In most cases (with Stargate) you can just play an audio response instead of involving your PC.

I find this preferable since playing *.wav* files means the Megacontroller must be running, and typically the reaction time from your PC is not going to be as immediate as Stargate's onboard audio responses.

To check out how this is done, highlight the "THEN" line in the schedule, go to the right side of the Megacontroller screen and select "Add." Select *"voice"*, *"play voice response"*, and you the get to choose from a myriad of stock selections. You can also record your own if you wish.

One last quirk about the sample schedule. Line 176 introduces an event that does Call Forwarding. I suspect that this is written for another device, (possibly the TimeCommander), because with Stargate you really don't need to use AT commands on an externally attached modem - *(i.e. you don't really need the modem)* - the "phone" features are an inherent part of Stargate.

My point is - that accomplishing the same thing is really easier that what you see here...

Give it a Try!

I think it's time you check out some of your skills (if you haven't already).

Go to *"File", "New",* and open a new schedule of your own. Keep in mind that a device database already exists that you can draw from (although you can add some of your own if you wish).

The first thing to do is go to the far right of your screen and click the yellow "new" button. This creates a new event on your schedule.

Give it a name, and check the other appropriate buttons for the type of event that you want, and off you go!

Actually, let me give you some exercises. Try creating the following events *(**before** checking out my sample solutions):*

1. Turn the family room lights on at 20% at 6 a.m. Monday thru Friday. Twenty minutes later they should go to full brightness.

2. Define a flag and name it "Home Mode." Write three separate events to do the following: When the security system is disarmed, the flag should be set (HINT - Digital Input "Sec. Armed" indicates state of security system - "ON" when system is armed). Secondly, the flag should be cleared when the security system has been armed for 5 minutes. And thirdly,... well, with the understanding that "Home Mode" is for when you're home - ***you*** decide. What should happen when the flag is set?

3. See if you can figure this one (you may have to define something): Make it so that any outgoing long-distance calls are blocked unless a code of your choosing is entered. Good luck!

Don't peek until you've tried your own hand at this, but ... when you have - mine are below:

```
2:
3:     EVENT: 1.
4:       If
5:         Time is 6:00 AM .MTWTF.
6:       Then
7:         (X:Family Room Lts.  A-1) Set Level  20 %
8:         DELAY  0:20:00
9:         (X:Family Room Lts.  A-1) Set Level  100%
10:      End
11:
12:    EVENT:  home flag set
13:      If
14:        (DI:Sec. Armed) Goes OFF
15:      Then
16:        (F:home)  SET
17:      End
18:
19:    EVENT:  home flag cleared
20:      If
21:        (DI:Sec. Armed) Goes ON
22:      Then
23:        DELAY  0:05:00 Re-Triggerable
24:        (F:home)  CLEAR
25:      End
26:
```

```
26:
27: EVENT:  home flag routine
28:        If
29:         (F:home) is SET
30:        Then
31:        ┌ If
32:        │   Before Sunrise  SMTWTFS
33:        │     -OR-
34:        │   After Sunset  SMTWTFS
35:        │  Then
36:        │   (X:Hallway Light  E-1) ON
37:        │   (X:Family Room TV  A-7) ON
38:        └ End
39:         (HVAC:Temp Zone 1)  Set Temp to 69 degrees
40:        End
41:
42: EVENT:  long-distance permit
43:        If
44:         TelePhone Seq:'^9876'  Received within 6 seconds
45:        Then
46:         (F:longdistance)  SET
47:         DELAY  0:00:10 Re-Triggerable
48:         (F:longdistance)  CLEAR
49:        End
50:
51: EVENT:  long-distance block
52:        If
53:         TelePhone Seq:'^1???????'  Received within 20 seconds
54:         and  (F:longdistance) is CLEAR
55:        Then
56:         Go ON Hook
57:        End
58
```

Chances are your solutions won't look exactly like mine. You'll soon find that there are a number of ways of doing the same thing.

But sometimes one way may be more efficient than others - or it may just work better for some (seemingly) unknown reason.

When schedules get complex and lengthy, I've found that it's better to keep your events somewhat short. By that I mean that you'll sometimes be better off breaking one long event into two or more. Stargate isn't terribly quirky, but it seems to prefer chewing on a multitude of smaller events rather than a few lengthy ones.

You can continue to play around with this for as long as you want. If you want you can also run a *"rules check"* to verify that you don't have any blatant errors in your schedule. You'll find this under *"File", "Rules Check."* This doesn't guarantee that your events will go as planned, but it *will* help avoid "brain-dead" mistakes, like leaving a line blank.

Good Luck with Playing!

When you're done with that, you're ready to dive right in ....

Part Three:

Finally! Putting it Together!

There's really only one thing you want to plan for:

... Everything!

I know that sounds kind of dramatic, but what I mean is that it's much easier to run *more* wire than you need *while you're already running it* -

... than to come back later and run it a second time.

Having said that, let me make some ***recommendations*** on what you should do:

- ***Pick a central location where ALL of your equipment can interface, and run all your cabling from there. Audio, Video, Stargate, Security - everything should be located here, or at least be easily accessible in the future from here. The most likely exception would be your Home Theater equipment, but make sure that you pull plenty of extra Cat5 & coax to the Theater Room.***

- ***Make individual cable runs for each phone jack using Cat5. Home-run rather than loop them.***

- ***Do the same thing for data jacks - even if you don't plan to use them right now.***

- ***If there's any possibility you might add the Stargate keypad, run a Cat5 RS485 loop through all possible locations, including the furnace rooms in case you want RS485 control over HVAC.***

- ***DO NOT FORGET to run a Cat5 cable from your PC to the Stargate. All your programming and schedule downloading must be done from your PC.***

Let me expand a little on some back issues of a couple of newsletters you may have received from me. First, a little about **wiring in new construction.**

Prewiring:

Obviously new construction is a lot easier to get cable where you want than in an existing house, but you have to watch for a few things:

First of all, you should be aware of ***building codes*** in your area. They often vary from county to county. You can probably obtain the information you need from your builder, or if you know someone who does this type of thing for a living, they can be a helpful resource, too. Naturally, you can always go to your county officials, & hopefully get straight answers.

If you don't know anyone in a low-voltage industry like security or telephone, then it might be good to **strike up a relationship with a local wire distributor who may be savvy to local code.**

Hey, this is a good thing anyway. If you can **get set up with a house account and buy your materials from them,** you can save substantially over what it would cost to buy from retail.

Generally speaking, though, here are a few tips: (check these out with County Code in your area):

1. TOOLS: Gosh. You need these. Actually, you don't need too much. You'll need a drill with wood "spade" bits (if you can get a cordless you'll thank yourself - I'd recommend 18V or better).

You'll also need a stepladder, plastic "tie wraps" (ask a cabling distributor for these - or Radio Shack if you've got mucho moolah), and either a staple gun or some kind of wire anchors for hanging your cabling.

NOTE from Experience: If you use a staple gun, BE CAREFUL! LOOK at every staple you fire! It's worth the extra time to verify that you didn't shoot through the cable when you consider the aggravation that could come from trying to troubleshoot later. Also be sure that you don't "pinch" the cable too tightly.

2. TRICKS: While you want to keep your wire runs up inside the joists so they will be out of the way of drywall, you can take advantage of water pipes and HVAC ductwork by following them. This will save you a lot of drilling through ceiling joists. Always look for the simplest path!

3. Low-voltage wire usually cannot be run vertically through cold-air returns, but it might be permissible to run horizontally through the chase. Check to see whether you'll need to enclose it in conduit. In commercial construction, your wire would also have to be "plenum" grade, though I personally haven't found this to be an issue residentially.

4. Wait until all the other mechanicals have been completed (electric, HVAC, etc.) to do your runs. I can't tell you how many times I've had to repair my cabling because someone else butchered it with a drill - or just plain cut it because it was "in their way." Of course, having said that, make sure you don't do the same thing. Always check the other side of a stud or wall before you drill. Nothing quite like hammer-drilling through a breaker box!

5. Drill your own holes through studs - don't use the electrician's! You want to keep as far away from his stuff as is practical so that you don't pick up electrical noise.

6. When you're done, you might need to fill the extra space in the holes you drilled with "firestop" (depending on code), which can be found at just about any hardware store. This rule usually only applies to holes drilled vertically inside of walls (through floors and ceilings).

7. Plastic Nail-up boxes: For volume controls, etc. you'll either need to install extra deep boxes, or cut the backs off with a hacksaw (which is what I do). You can also buy specialty backless boxes at some crazy cost of several dollars apiece.

Personally, I'd rather spend 39 cents & cut the backs off...

...a retrofit box with "wings" to clamp on existing drywall.

Postwiring:

If, on the other hand, your project requires you to run cabling in **existing construction**, you've got more of a challenge:

Obviously, this is a little different than in new construction. The bad news is that it can be a little trickier - the good news is that you don't have to worry about someone else hacking your work after you leave the job site!

The way you'll approach pulling wire from room to room depends on how your home is built. If you have a one story house built on a slab, your likely route will be up inside the wall and through the attic. If you have a crawl space or unfinished basement you would naturally take that route.

Let's first examine how you pass wiring inside of walls without destroying stuff. Here are the tools that you need:

> **Fish Tape.** This is something that you can find at any decent hardware store. If you're unfamiliar with it, it's simply a coil of flattened steel wire. You can unroll it

to pull or push wire through ceiling or wall spaces.

Stud Finder. Same thing. Hardware store. This helps you locate studs in your wall so you don't cut holes in drywall where you don't mean to!

Miscellaneous. Drywall knife, Retro-fit (wing-back) electrical boxes, Electrical tape, flashlight, measuring tape,... stuff you'll find in your junk drawers (you *do* have junk drawers, don't you?).

Let's assume for the moment you want to go up the wall into the attic. You'll be cutting a hole into the drywall where the wires will exit into the room. Using your stud finder (guys at the hardware store can explain this device to you), locate the studs in your wall and trace the outline of your retrofit electrical box with a pencil (someplace where the studs are NOT).

OH - BY THE WAY: *Scope out your path BEFORE you do anything!* ***First,*** *make sure there are no obstructions in the wall with your stud finder. If you don't detect anything this way, then cut just a small hole and run your fish tape up the wall to make sure it doesn't hit anything before it gets to ceiling height.*

Also, if you're cutting into ***wallpaper****, you'll want to score it with a knife before you cut with your drywall saw - and make your cut on the INSIDE of the scoring. This way you won't tear the paper.*

If by chance your location turns out to be a mistake, you have a couple of choices: either repair the drywall, or (if it's wallpaper) cover it with a blank plate & beg your spouse's forgiveness (do some extra chores or something - only a scoundrel would hide it with the big comfy chair).

Second, measure the exact distance of your proposed cut from a corner, & go up into the attic & do the same. You need to be able to drill down into the wall chase at the *exact* spot above your proposed "hole-in-the-wall" in the room below.

While you're in the attic (if everything looks clear), drill your hole and drop your fish tape all the way down (if there's no insulation you can just drop a weighted string). You're now ready to cut the drywall below, reach in to grab the string or fish tape, use electrical tape to tie your wiring on, and go back up into the attic to pull away!

At the other end of your run, you'll repeat the process. The order in which you do things may vary, but this is basically the how it's done.

Not too difficult, was it? Except for the nasty attic insulation (ecch!)....

IF YOU'RE RUNNING THROUGH THE BASEMENT it's not much different. Again you'll measure carefully, scope out the path... but this time you'll be drilling UP from the basement into the wall space.

To make sure you drill up into a wall space and not through your hardwood floor (been there done that), measure carefully off some reference point visible from both the basement and upstairs. Heat ducts make good reference points.

You can also look for nails and water pipes running up from the basement as good indicators of wall chase locations. If you want to play it safe, drill a small pilot hole first and check that it's in the right place.

IF YOU WANT TO DRILL THROUGH A CARPETED FLOOR there's a right way and a wrong way. The wrong way is to drill directly through the carpet. You'll watch in amazement as your beautiful carpet rolls up on the drill bit.

What you really want to do is cut an "X" in the carpet with your utility knife, and pull the folds back with one hand while you slowly drill with the other. If you use a good sharp knife, the carpet will fold back down nicely & the cuts will be just about invisible.

ONE OTHER CAUTION: Do your best to know where electrical runs are inside the walls when you drill. Blue flame is pretty, but I think you'd rather not see it. Remember that romex (electrical cable) often runs directly attached to the vertical studs, though not always - and that it may run horizontally from outlet to outlet. Don't be paranoid, just be careful. **Drill slowly when you think you're about to break through, and once your drill exits the other side of a wood stud or plate, BACK OFF.**

FROM BASEMENT TO SECOND FLOOR: Now you *have* to be clever. Hypothetically speaking, you could violate code and find a cold air return that runs directly from the basement in a straight line to the second floor. If so, you'd just have to pop open the return in the basement, and open the vent on the second floor so you could drill into the attic space.

Aside from this, you might find a common chase that was built into the house for pipes or cabling, etc. Sometimes a chimney has dead space around it that you can use. If you can't find a good hidden path, running wires through closet spaces or laundry chutes may accomplish the mission. If need be, you can then dress it up with "wire molding" from the hardware store so it's not so unsightly.

The last resort would be to take the wiring outside the house, hide it best as you can by running it behind downspouts/gutters, etc., & back into the basement/attic at the other end.

Review:

Now that you understand the basic ideas concerning how to interface with each subsystem of your home, you're ready to begin putting it all together.

As a quick review, let's consider again what we've covered:

- **X10:** No hardwiring required. You use existing electrical wiring to carry signals from a) Stargate, b) a base receiver that receives UHF commands, c) an X10 outlet, or d) some other X10-based controller. Signal bridges/couplers/repeaters can help with problems passing X10 commands.

- **UHF:** Absolutely simple to install with no wiring requirements. Just watch your distance limitations, & bear in mind that, depending on your application, it might prove to be a little more difficult to interface with.

- **IR:** You can use an IR-to-UHF-back-to-IR device if you don't want to run wire. If you want the dependability of a hardwired system, use either shielded cable (3 conductors minimum) or Category 5.

- **Audio:** 16 gauge stranded & twisted speaker wire at least (14 is better). Four-conductors from the home run location to each volume control, & a two-conductor to each of two speakers per volume control. You can pull your IR wire right along with your four-conductors to save effort. Give thoughtful attention to speaker placement.

- **Video:** Run all video sources that you want distributed through a modulator. From the modulator, use a "combiner" to inject your sources back into the Cable system.

- **Theater:** Be sure to pull plenty of coax and Cat5 to your equipment if its location is different from your "brain room." Powered subwoofers are best for Theater; passive subs are generally OK for music. Choose between Dolby Digital (AC-3) or Dolby Pro-logic (remember that you need *digital* sources (like DVD's) to enjoy the advantages of AC-3.

- **HVAC:** Can be controlled either with X10 based or hardwired smart thermostats.

- **Security:** Hardwired sensors are easiest to interface with. You just need a basic system (as far as automation features are concerned). The wiring from each sensor that you want Stargate to monitor must pass through its own dedicated digital input on Stargate.

- **Stargate:** Once you have everything connected and visible to Stargate, you just have to write your events!

There's no need for me to parrot everything that's in Stargate's manual as far as installation is concerned, and your other hardware will also have its own unique sets of directions.

Once you have your physical infrastructure in place *(i.e. your X10 devices, IR system, thermostats, & other low-voltage thingies)* you're ready to begin designing your schedule of events. Per Stargate's manual, you connect your PC via its serial port to the COM1 port on the Stargate panel. Open the "WinEvm" software *(that's Windows Event Manager)*, and verify that you can open the "Megacontroller" window. This simply demonstrates communication between your PC and Stargate. If you have trouble, see the Stargate manual for trouble-shooting remedies. Otherwise, if all is well, you're ready to go!

Let me just encourage you again: if you're reading this for the first time and haven't tried your hand at any of it yet ...

Don't be Afraid!

If you're completely new and don't want to invest a lot of money to begin, start out with some *X10* devices. You can get some quick experience and gain confidence in your abilities this way without committing to bigger budget items.

Once you've got a grip on that, you really should consider getting the Stargate, Timecommander Plus, or something else that can perform the intelligent functions of home automation.

From here, it's just a matter of adding on when you want to expand your system! You can develop your script writing skills as you go along, and pretty soon you'll be surprising even yourself.

This just about concludes **Book One** of **Integrating the Smart Home and its Owner**. You've already done more to learn about home automation than most people ever do. But obviously there's more ...

And now, the absolute best thing you can do is **dive right in**. Remember, you don't have to know everything (or even all that much) to start. So much of what you're about to learn comes by ***doing!*** "The water's fine" so jump in - you'll soon be on your way to mastering a skill (an art?) that will provide you with tremendous satisfaction for years to come.

The second half of this book (Book 2) will walk you right through a real-life installation of the hardware & software I've discussed here.

Meanwhile, I'll leave you with some scripts and diagrams that will help demonstrate just some of the things you can do with Stargate. The diagrams are straight from Stargate's manual *(in fact, a lot of what follows is available in more complete form from the JDS website)* .

The first scripts you see are direct from the JDS *appnotes.pdf* which you can download for yourself (www.jdstechnologies.com/download/appnotes.pdf). I've included just some of them here as a convenience for you - following then are some scripts of my own.

Appendix

Diagrams:

This is what the Stargate panel looks like. It gives you an idea of the layout of digital inputs, relays, analog inputs, etc. Doorbells, security sensors, telephone lines, & other low-voltage devices that you want to monitor can be passed through the digital inputs. The relays can be used for a variety of uses, such as turning speakers on/off to individual rooms. Analog inputs can be used for relatively inexpensive temperature sensors.

STARGATE Specifications

STARGATE Panel

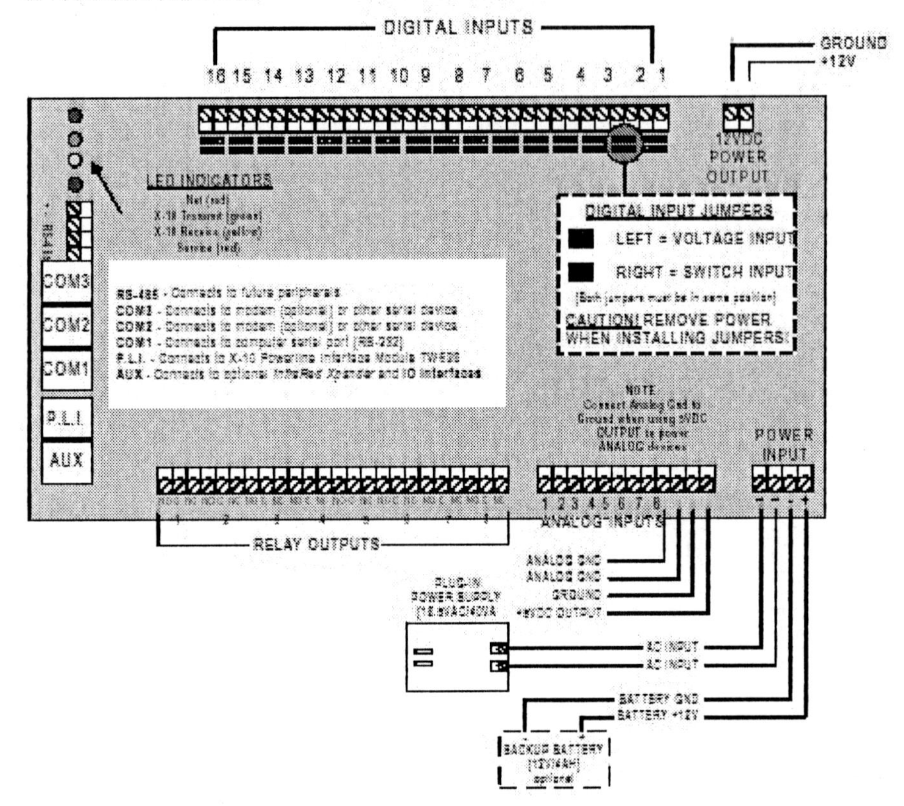

The JDS IR-Xpander is useful for automating your IR System. You can run the emitters directly to your components or, if you already have a whole-house IR system (Niles, Xantech, Jamo), you can simply run a single emitter to one of your IR sensors. Training the IR-Xpander is much like any other learning remote (it learns directly from your existing remotes).

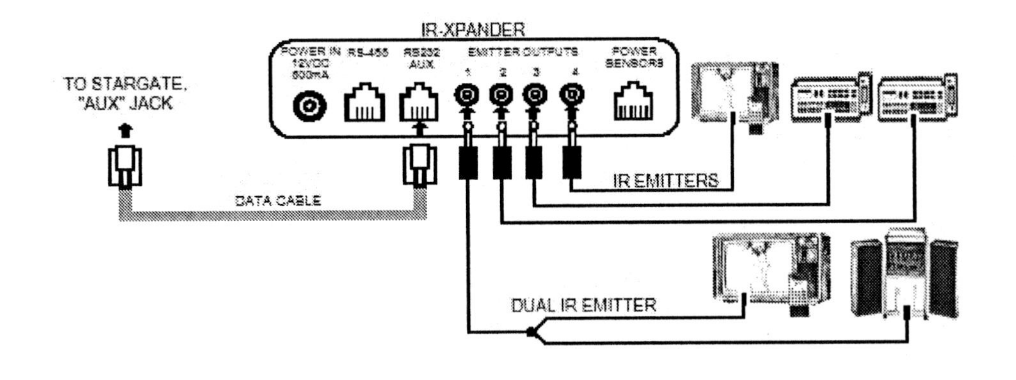

You can "daisy-chain" up to 16 LCD-96M keypads to Stargate's RS485 port with two twisted pairs of Cat5 cabling. Up to four of them can be powered by Stargate's onboard power (more than that requires an external power supply). They can be used to control security, a/v, voice-mail, HVAC - I guess just about anything that Stargate can do.

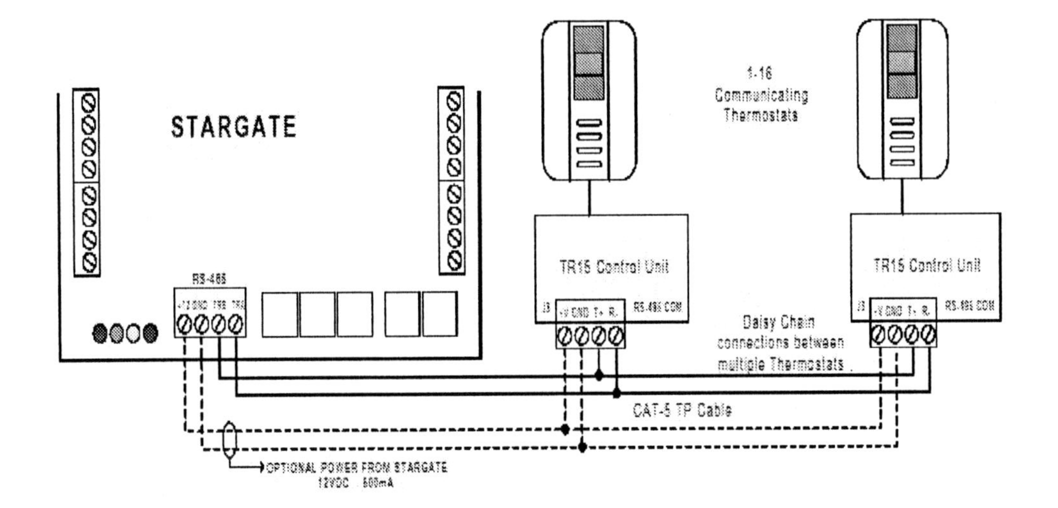

Scripts:

The first scripts you see below are directly from Stargate's manual.

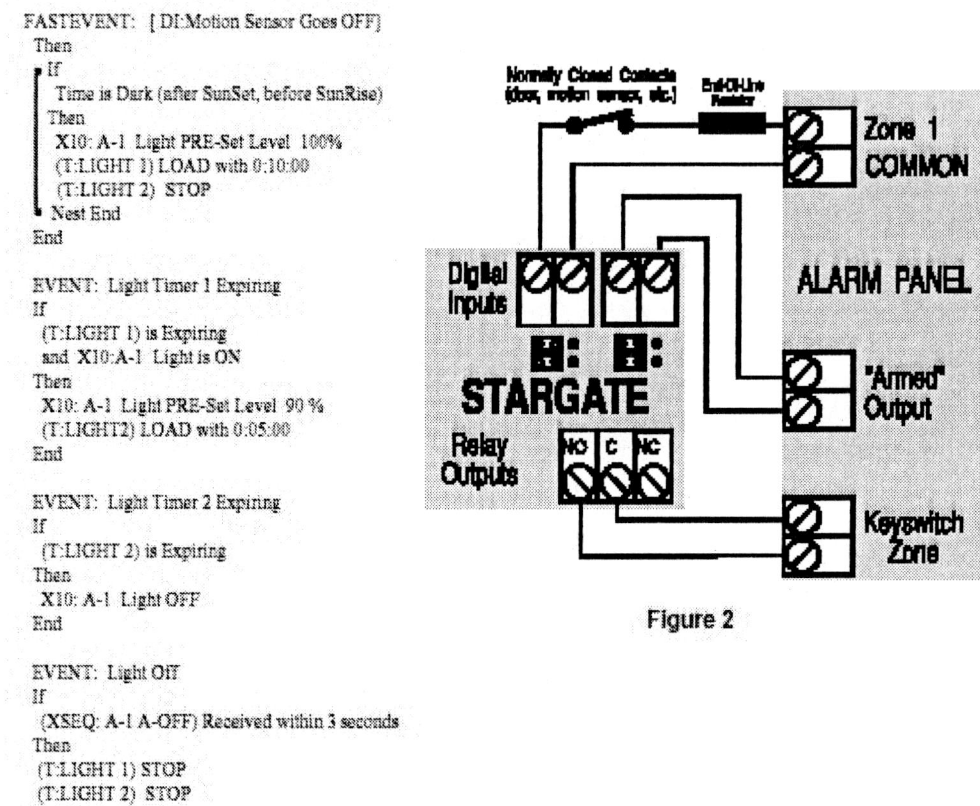

```
FASTEVENT:  [ DI:Motion Sensor Goes OFF]
 Then
  If
   Time is Dark (after SunSet, before SunRise)
  Then
   X10: A-1  Light PRE-Set Level  100%
   (T:LIGHT 1) LOAD with 0:10:00
   (T:LIGHT 2)  STOP
  Nest End
 End

 EVENT:  Light Timer 1 Expiring
 If
  (T:LIGHT 1) is Expiring
  and  X10:A-1  Light is ON
 Then
  X10: A-1  Light PRE-Set Level  90 %
  (T:LIGHT2) LOAD with 0:05:00
 End

 EVENT:  Light Timer 2 Expiring
 If
  (T:LIGHT 2) is Expiring
 Then
  X10: A-1  Light OFF
 End

 EVENT:  Light Off
 If
  (XSEQ: A-1 A-OFF) Received within 3 seconds
 Then
  (T:LIGHT 1) STOP
  (T:LIGHT 2)  STOP
 End
```


Figure 2

Announcing Caller ID Through Audio System

The following events check the Caller ID of each incoming call and play a unique User Voice Response for each number in the Caller ID List events. Since there is a limit to the number of If - CallerID lines that can be placed in a single event (about 20), separate events have been created, each containing a portion of the CallerID numbers to be announced by a unique User Voice Response. Calls that are not included in a Caller ID List event will be announced by phone number only.

```
EVENT:  CALLER ID LIST 1
If
  CallerID: ???1111111
  or  CallerID: ???2222222
  or  CallerID: ???3333333
Then
  (F:CallerID)  SET
┌ If
│   CallerID: ???1111111
│  Then
│   Voice: Moe  [Line]
└  Nest End
┌ If
│   CallerID: 2222222222
│  Then
│   Voice: Larry  [Line]
└  Nest End
┌ If
│   CallerID: ???3333333
│  Then
│   Voice: Curly  [Line]
└  Nest End
End

EVENT:  CALLER ID LIST 2
If
  CallerID: ???4444444
  or  CallerID: ???5555555
  or  CallerID: ???6666666
Then
  (F:CallerID)  SET
┌ If
│   CallerID: ???4444444
│  Then
│   Voice: Peter  [Line]
└  Nest End
┌ If
│   CallerID: ???5555555
│  Then
│   Voice: Paul  [Line]
└  Nest End
┌ If
│   CallerID: ???6666666
│  Then
│   Voice: Mary  [Line]
└  Nest End
End
```

Managing CD's With LCD Keypad

The following two events will allow you to select cd's from the LCD Keypad and display the name of the selected disk on the cd menu. Since CD management is not a built-in function of the keypad, an event must be added to the schedule to manage cd selection. When a selection is made, the LCD Keypad loads the "CD" Variable with a value corresponding to the cd's slot number in the multiple disk player. For each CD Variable value, a nested if-then is created to issue the required IR commands and change the text on the cd menu to display the selection. The event then switches the keypad to the CD menu screen and loads the CD Variable back to 0.

```
EVENT: CD SELECT
 If
  (V:CD) > 0
 Then
 ┌If
 │ (V:CD) = 1
 │Then
 │ LCD: Change Text Line 1 Menu 18 to 'AbbeyRoad ' [KP:# 13]
 │ (IR:CD DISK ) play 1 time(s) [Emitter1]
 │ (IR:CD 1 ) play 1 time(s) [Emitter1]
 │ (IR:CD ENTER ) play 1 time(s) [Emitter1]
 └Nest End
 ┌If
 │ (V:CD) = 2
 │Then
 │ LCD: Change Text Line 1 Menu 18 to 'HardDaysNi' [KP:# 13]
 │ (IR:CD DISK ) play 1 time(s) [Emitter1]
 │ (IR:CD 2 ) play 1 time(s) [Emitter1]
 │ (IR:CD ENTER ) play 1 time(s) [Emitter1]
 └Nest End
 ┌If
 │ (V:CD) = 3
 │Then
 │ LCD: Change Text Line 1 Menu 18 to '  Help!  ' [KP:# 13]
 │ (IR:CD DISK ) play 1 time(s) [Emitter1]
 │ (IR:CD 3 ) play 1 time(s) [Emitter1]
 │ (IR:CD ENTER ) play 1 time(s) [Emitter1]
 └Nest End
 ┌If
 │ (V:CD) = 4
 │Then
 │ LCD: Change Text Line 1 Menu 18 to 'Let It Be ' [KP:# 13]
 │ (IR:CD DISK ) play 1 time(s) [Emitter1]
 │ (IR:CD 4 ) play 1 time(s) [Emitter1]
 │ (IR:CD ENTER ) play 1 time(s) [Emitter1]
 └Nest End
  LCD: Goto Menu Screen = Menu 18 [KP:# 13]
  (V:CD) LOAD with 0
 End
```

```
EVENT: HVAC*
 If
  TelePhone Seq:'4822*' Received within 6 seconds
 Then
  (F:HVAC) SET
  Voice:THE TEMPERAT IS Zone1 Temperature DEGREES SET Zone1 Setpoint DEGREES [CO,ICM]
  Voice:TO SET TEMPERAT PRESS 1 [CO,ICM]
  Voice:FOR COOL MODE PRESS C O [CO,ICM]
  Voice:FOR HEATING MODE PRESS H E [CO,ICM]
  Voice:FOR AUTOMATI MODE PRESS A U [CO,ICM]
  Voice:FOR OFF MODE PRESS 0 0 [CO,ICM]
  Voice:FOR FAN ON PRESS F A STAR [CO,ICM]
  Voice:FOR FAN OFF PRESS F A POUND [CO,ICM]
  Voice:TO LEAVE THIS MENU PRESS POUND 0 [CO,ICM]
 End

EVENT: HVAC SELECT
 If
  (F:HVAC) is SET
 Then
  If
   TelePhone Seq:'1' Received within 2 seconds
  Then
   Voice:ENTER TEMPERAT FOLLOWBY POUND [CO,ICM]
   TouchTone to user_VAR SYNC
   (HVAC:Zone1) Load Setpoint with value in user_VAR
   Clear TouchTone Input Buffer
  Nest End
  If
   TelePhone Seq:'26' Received within 3 seconds
  Then
   (HVAC:Zone1) COOL Mode
   Voice:COOL MODE [CO,ICM]
  Nest End
  If
   TelePhone Seq:'43' Received within 3 seconds
  Then
   (HVAC:Zone1) HEAT Mode
   Voice:HEATING MODE [CO,ICM]
  Nest End
  If
   TelePhone Seq:'28' Received within 3 seconds
  Then
   (HVAC:Zone1) AUTO Mode
   Voice:AUTOMATI MODE [CO,ICM]
  Nest End
  If
   TelePhone Seq:'00' Received within 3 seconds
  Then
   (HVAC:Zone1) OFF Mode
   Voice:OFF MODE [CO,ICM]
  Nest End
  If
   TelePhone Seq:'32*' Received within 3 seconds
  Then
   (HVAC:Zone1) Fan ON
   Voice:FAN ON [CO,ICM]
  Nest End
  If
   TelePhone Seq:'32#' Received within 3 seconds
  Then
   (HVAC:Zone1) Fan OFF
   Voice:FAN OFF [CO,ICM]
  Nest End
  If
   TelePhone Seq:'#0' Received within 3 seconds
   or TelePhone Seq:'+' Received within 2 seconds
  Then
   (F:HVAC) CLEAR
   Voice:BDRINGGG [CO,ICM]
  Nest End
 End
```

DEFAULT INTERCOM

In many cases it is more convenient to have Stargate default to the INTERCOM rather than the CO line. Selecting ICM under Define - Telephone Setup - Phone Output Default will cause all phones to be connected to the ICM port by default. Dialing "9" will switch the user to the outside line (CO). When this setup is used, it is important to note that all phones connected to Stargate's "PHONE" port will not ring since they are normally not connected to the CO line. The following events will switch Stargate's "PHONE" port to the CO line when the line rings and switch back to the ICM line when the phone hangs up after a call. Additional events are used to accomodate abandoned calls (the caller hangs up before you or Stargate answers the call). The OFF-HOOK AUTO SWITCH event (optional) automatically switches an off-hook phone to the CO line after 3 seconds if no touchtone is pressed. This is useful for those who are not familiar with the system and don't know to dial "9" for the outside line. If any touchtone other than "9" is pressed after going off-hook, Stargate remains connected to the ICM line.

```
EVENT: DEFAULT ICM RING
If
  CO: Ring 1
  or CO: Ring 2
  or CO: Ring 3
  or CO: Ring 4
Then
  Connect PHONE port to CO port
  (T:RING) LOAD with 0:00:08
End

EVENT: DEFAULT ICM RESET
If
  TelePhone Seq:'+' Received within 2 seconds
Then
  DELAY 0:00:02
 ┌If
 │ CO: Is ON Hook
 │Then
 │ Connect PHONE port to ICM port
 └Nest End
End

EVENT: ABANDONED CALL
If
  (T:RING) is Expiring
  and CO: Is ON Hook
Then
  Connect PHONE port to ICM port
End

EVENT: OFF-HOOK - AUTO SWITCH
If
  (T:RING) is Not Running
Then
 ┌If
 │ ICM: Is OFF Hook
 │Then
 │ (T:DIAL TONE) LOAD with 0:00:03
 └Nest End
 ┌If
 │ TelePhone Seq:'^?' Received within 4 seconds
 │Then
 │ (T:DIAL TONE) STOP
 └Nest End
 ┌If
 │ (T:DIAL TONE) is Expiring
 │Then
 │ Connect PHONE port to CO port
 └Nest End
 ┌If
 │ ICM: Is ON Hook
 │ and (T:DIAL TONE) is Running
 │Then
 │ (T:DIAL TONE) STOP
 └Nest End
End
```

Following are some of my own scripts which I'll try to explain. The first one sets a flag when a new message is received in voicemail. Then, when someone enters the room Stargate announces the number of voice-mail messages over the stereo speakers. If the stereo is playing, a macro switches the relays to connect the speakers to Stargate. Finally, the flag is cleared so that the message doesn't play again (don't mind the setting of the last flag - it's related to another event).

```
EVENT: message notify
If
 (VMAIL:MBX-1 # New Msg)  Increases in Value
Then
 (F:annouce messages)  SET
End

EVENT: message notify 2
If
 (F:annouce messages) is SET
 and (DI:family PIR) Toggles
Then
  If
   (DI:stereo pwr) is ON
  Then
   (THEN MACRO:SG spkr switch)
   DELAY 0:00:07
   (THEN MACRO:Ster.spkr switch)
  Nest End
 DELAY 0:00:01
 Voice:STTRHAIL [Spkr]
 VM:Say number of Messages in MailBox 1 [Spkr]
 (F:annouce messages) CLEAR
 (F:annc msgs again) SET
End
```

This next one lets me set a wake-up alarm from my bedside phone. The code "666" switches the telephone to Intercom mode (doing away with dial tone so I can hear) & announces the existing alarm time. "Touchtone to Time Label" is a function that equates telephone keypad presses with clock time (I use "*" for a.m., "#" for p.m.). Doing so sets another flag so Stargate will perform the function(s) I tell it to do when the alarm time arrives. That could be to turn on lights, start coffee, turn up the heat, play soft music - or in my case, just play a macro which yells at me until I get up. Once I'm out of bed, the motion sensor detects movement & stops the routine.

```
EVENT:  wakeup set
If
  TelePhone Seq:'*666' Received within 5 seconds
Then
  Connect PHONE port to ICM port
  (F:intercom)  SET
  Voice:WAKE_UP <wakeup>  [ICM]
  TouchTone to Time Label:(TL:wakeup)
End

EVENT:  clear intercom
If
  (F:intercom) is SET
  and  ICM: Is ON Hook
Then
  Connect PHONE port to CO port
  (F:intercom)  CLEAR
End

EVENT:  wakeup time converted
If
  TouchTone to TimeLabel complete
Then
  Voice:WAKE_UP TIME <wakeup>  [ICM]
  (F:set wakeup)  SET
End

EVENT:  waking up
If
  Time is (TL:wakeup)
  and  (F:set wakeup) is SET
Then
  (THEN MACRO:wakeup routine)
End

EVENT:  awake
If
  (T:gett'n out'a bed) is Running
Then
  If
    (DI:family PIR) Toggles
  Then
    (T:gett'n out'a bed)  CLEAR
    Voice:YOUDAMAN  [Spkr,ICM]
  Nest End
End
```

If you'd like to know when the kids get home, the following script will let you do just that. You set a flag when you leave home to enable the routine. It gives you five minutes to set the alarm and leave before it goes into effect. Then when someone arrives home and disarms the security system, you can receive a phone call or page.

```
EVENT:  set page if entry
 If
  TelePhone Seq:'555' Received within 7 seconds
 Then
  " set flag to notify when someone "
  " arrives home "
  (F:page if entry)  SET
  DELAY  0:05:00
 ┌ If - Always
 │  (DI:sec.armed) is OFF
 │ Then
 │  (F:page if entry)  CLEAR
 │  " flag is only set if security  "
 │  " system is armed "
 └ Nest End
 End

EVENT:
 If
  " when flag is set and someone  "
  " disarms security, call me! "
  and  (DI:sec.armed) Goes OFF
  and  (F:page if entry) is SET
 Then
  (F:page if entry)  CLEAR
  TelePhone Out:'*5555555,,,'
  Voice:SOMEONE IS HOME  [CO,ICM]
  DELAY  0:00:02
  Voice:SOMEONE IS HOME  [CO,ICM]
  DELAY  0:00:02
  Voice:SOMEONE IS HOME  [CO,ICM]
  DELAY  0:00:01
  Voice:SOMEONE IS HOME  [CO,ICM]
  DELAY  0:00:02
  Voice:SOMEONE IS HOME  [CO,ICM]
  DELAY  0:00:01
  Voice:SOMEONE IS HOME  [CO,ICM]
  TelePhone Out:',,+'
  (F:AlarmAcknowledge)  CLEAR
  DELAY  0:00:01
  TelePhone Out:',,+'
  DELAY  0:00:02
 End
```

This is just a random light routine that runs when an AWAY flag is set (you can enable this however you want - by arming security, by phone keypad presses, etc.). The THEN MACRO that's referenced follows the first two events. Every so many minutes, Stargate checks to see if the Random Lighting or Home flag is still set. If not, the routine continues. At the end of the macro, the Random Lighting flag goes into and "idle" state and then back into a "set" state. This is so that Stargate will detect a change in state and repeat the event.

```
EVENT: random lights2
If
 (F:away) is SET
 and After Sunset SMTWTFS
   -OR-
 (F:away) is SET
 and Time is Before 1:00 AM SMTWTFS
Then
 (F:random lighting) SET
Else
 (F:random lighting) CLEAR
End
```

```
EVENT: random lights=then macro
If
 (F:random lighting) is SET
Then
 (THEN MACRO:random lights)
End
```

```
Macro:random lights
 1:    MACRO BEGIN
 2:     DELAY 0:00:03
 3:     If
 4:       (F:home) is Not SET
 5:       and (F:random lighting) is Not CLEAR
 6:     Then
 7:       X10: B-1 kitchen OFF
 8:       X10: B-4 kitchen chandelr PRE-Set Level 74 %
 9:       X10: B-3 front porch PRE-Set Level 90 %
10:     Nest End
11:     DELAY 0:19:00
12:     If
13:       (F:home) is Not SET
14:       and (F:random lighting) is Not CLEAR
15:     Then
16:       X10: B-4 kitchen chandelr OFF
17:       X10: B-1 kitchen PRE-Set Level 77 %
18:       X10: B-3 front porch PRE-Set Level 52 %
19:       X10: B-8 study lamp Set Level 50 %
20:     Nest End
21:     DELAY 0:11:00
22:     If
23:       (F:home) is Not SET
24:       and (F:random lighting) is Not CLEAR
25:     Then
26:       X10: B-8 study lamp OFF
27:       X10: B-2 garage flouresc ON
28:        DELAY 0:00:17
29:       X10: B-2 garage flouresc OFF
30:     Nest End
31:     DELAY 0:33:00
```

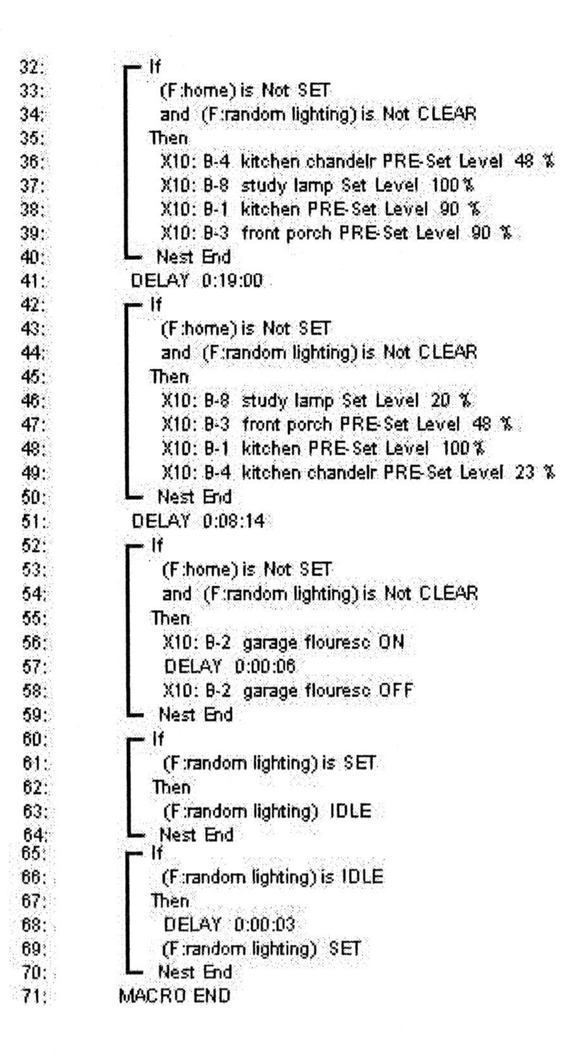

```
32:        If
33:          (F:home) is Not SET
34:          and (F:random lighting) is Not CLEAR
35:        Then
36:          X10: B-4 kitchen chandelr PRE-Set Level 48 %
37:          X10: B-8 study lamp Set Level 100 %
38:          X10: B-1 kitchen PRE-Set Level 90 %
39:          X10: B-3 front porch PRE-Set Level 90 %
40:        Nest End
41:      DELAY 0:19:00
42:        If
43:          (F:home) is Not SET
44:          and (F:random lighting) is Not CLEAR
45:        Then
46:          X10: B-8 study lamp Set Level 20 %
47:          X10: B-3 front porch PRE-Set Level 48 %
48:          X10: B-1 kitchen PRE-Set Level 100 %
49:          X10: B-4 kitchen chandelr PRE-Set Level 23 %
50:        Nest End
51:      DELAY 0:08:14
52:        If
53:          (F:home) is Not SET
54:          and (F:random lighting) is Not CLEAR
55:        Then
56:          X10: B-2 garage flouresc ON
57:          DELAY 0:00:06
58:          X10: B-2 garage flouresc OFF
59:        Nest End
60:        If
61:          (F:random lighting) is SET
62:        Then
63:          (F:random lighting) IDLE
64:        Nest End
65:        If
66:          (F:random lighting) is IDLE
67:        Then
68:          DELAY 0:00:03
69:          (F:random lighting) SET
70:        Nest End
71:      MACRO END
```

This is a little reminder to myself to take the trash out on Sunday nights. If it's after 8 p.m. and a motion sensor detects movement in the family room, Stargate plays a little voice reminder. The routine repeats every hour (waiting for my presence each time so that it knows I'm close enough to the speakers to hear it). Incidentally, I inserted a similar routine inside this event regarding a doctor's appointment. However, I just now noticed that the message was played over the telephone (CO). *No wonder I missed that appointment!*

```
EVENT: trash night
If - Always
  Time is After 8:00 PM S.....
  and (DI:family PIR) Toggles
Then
  Voice:trash night [Spkr]
    If
      Date is After Mar 15
      and Date is Before Mar 27
    Then
      Voice:dr.grote appt. [CO]
    Nest End
  DELAY 1:00:00
End
```

Here's an If-Always event for when the security alarm is triggered. All lights that are on House Code B flash on and off every 2 seconds for as long as the Alarm flag is set. In an event like this, some sort of delay (2 seconds in this case) is necessary to allow any other X10 signals to execute. If there was no delay, the powerline would be completely flooded with the repeated X10 commands of this event.

```
EVENT: alarm lights
 If - Always
  (F:alarm) is SET
 Then
  X10: B - All Lights ON
  DELAY 0:00:02
  X10: B - All Lights OFF
  X10: B - All Units OFF
  DELAY 0:00:02
 End
```

The idea behind this event is two-fold. One is to turn the lights on automatically upon entrance to the house if it's dark. The other is to remind a certain someone to remove his dirty shoes when he comes in. The challenge to this event was that I only wanted it to execute when someone was *entering* the house, not *leaving*. The solution to the problem was to link it to movement inside. If the motion sensor sees movement within a 10 second time-frame prior to the door being opened, it does *not* execute the event. That would indicate that either someone is already home, or else is just leaving.

```
EVENT: entry front
 If
  (DI:family PIR) is ON
 Then
  DELAY 0:00:10 Re-Triggerable
   If
    (DI:front door) Goes OFF
    and (F:it's light) is CLEAR
   Then
    X10: B-4 kitchen chandelr PRE-Set Level 94 %
     If
      (DI:stereo pwr) is ON
     Then
      (THEN MACRO:SG spkr switch)
      DELAY 0:00:03
      (THEN MACRO:Ster.spkr switch)
     Nest End
    DELAY 0:00:01
    Voice:takeoffshoes [Spkr,ICM]
   Nest End
 End
```

BookII (Installation): First Things First...

Getting started in any sizeable project is always the hardest part. I recently read a comment on the JDSUsers site (www.jdsusers.com) from someone who seemed overwhelmed by the prospect of actually using their Stargate.

I can't stress enough that you don't really have to know exactly everything that you're going to do with your system before you begin. All you need to initially concentrate on is getting the infrastructure in place. You'll find that your system's functions will evolve with time and creativity.

As a matter of fact, as I sit here and begin this book about my own home installation, I only have a few notions about what my final design will be. If you haven't skipped ahead, your guess is as good as mine as to what you're going to read.

As an aspiring author I suppose that's a poor confession to make, but then again, it's not my intent to produce a literary masterpiece; rather to give you a hands-on demonstration that reflects reality.

Right now I should probably explain a few things about what you're getting into.

First of all, you may or may not know that this is more or less a sequel to **Integrating the Smart Home... (Volume 1)**. If you don't happen to have it, you should be aware that there's a lot of foundational information in that first book that I'm not going to cover here. Basically, it covers an overall view of a Stargate-based system with numerous 3rd party add-ons.

X10, infra-red systems, RCS thermostats, lighting, etc.... If you feel like you have a decent grasp of these things already, you're probably OK without the first edition.

If, however, you're a newbie, I'd really recommend that you get a hold of that first book before diving too much deeper. It really isn't terribly long because I tried to keep it to-the-point. At any rate, it will help to familiarize you with some of the items I'll be using in this installation.

Oh, by the way (guess I've been assuming too much): If you haven't read the first book or visited my website (www.Integratorpro.com), you probably don't know who the heck I am. And so for a brief bio, assuming (too much?) that you'd like to know:

My name is Andy Jackson and I run a small custom electronics business in the Cincinnati, Ohio area. It's actually the second time I've done this. Several years ago I presided over an S-corp with a partner (same line of work). I eventually sold my interest & worked for a few years as a network engineer for some local companies.

Currently I hold a Microsoft MCSE and Cisco CCNP ... but it will probably never matter. I just plain don't like being someone else's employee.

So let's get on with things....

***ALSO: LEGAL STUFF** - I can't/ won't/ don't accept any responsibility for what happens in your installation(s) as a consequence of reading this book. This book is written as an illustration of work that I did in my own home. Though I hope you find a lot of useful information here, the book's primary purpose is NOT to tell you how to do the same. If you have any question concerning code or safety for your own applications, you should consult a local professional.*

prohibited and will cause problems for both of us. So please!!

Wiring and Cabling:

This is the MAIN THING, especially if you're doing new construction. As long as you have wiring running all over the house, you'll be set to do whatever you might want to do at a later date.

The key to this is simple. After you've run wire for as many things as you can think of ...

... Run more. Coil extra runs in the attic or wherever seems to make sense (coax, cat5, 4-conductor speaker wire, etc.)

It almost never fails: *something* will call for a wire run down the road, and you'll find yourself either having to retro the job or find a way to do it wireless.

As an example, I spent three days just pulling wire in my home as it was being built. I hard-wired security sensors on every door and every window in the place - and forgot to run a cable for the keypad.

Oh well - it's a good thing I know *(sic)* what I'm doing.

I'll show you in a little bit how to fish that cable down a wall after drywall has gone up.

But for now, I want to begin by showing you the floor plan and giving you my initial ideas for the system design. The home isn't really that large (less than 3000 square ft.), but that's not terribly relevant anyway.

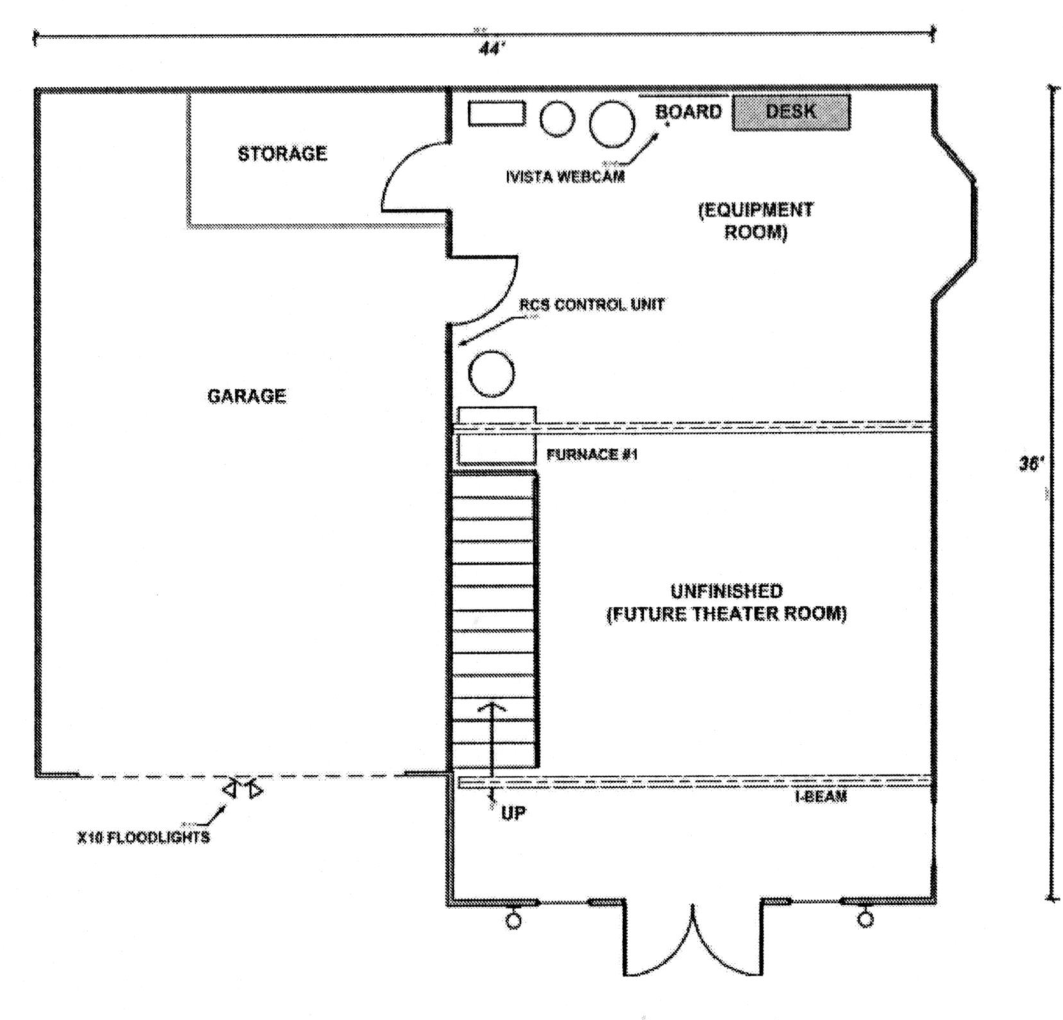
44'
STORAGE
BOARD
DESK
IVISTA WEBCAM
(EQUIPMENT ROOM)
RCS CONTROL UNIT
GARAGE
FURNACE #1
36'
UNFINISHED (FUTURE THEATER ROOM)
I-BEAM
UP
X10 FLOODLIGHTS

basement

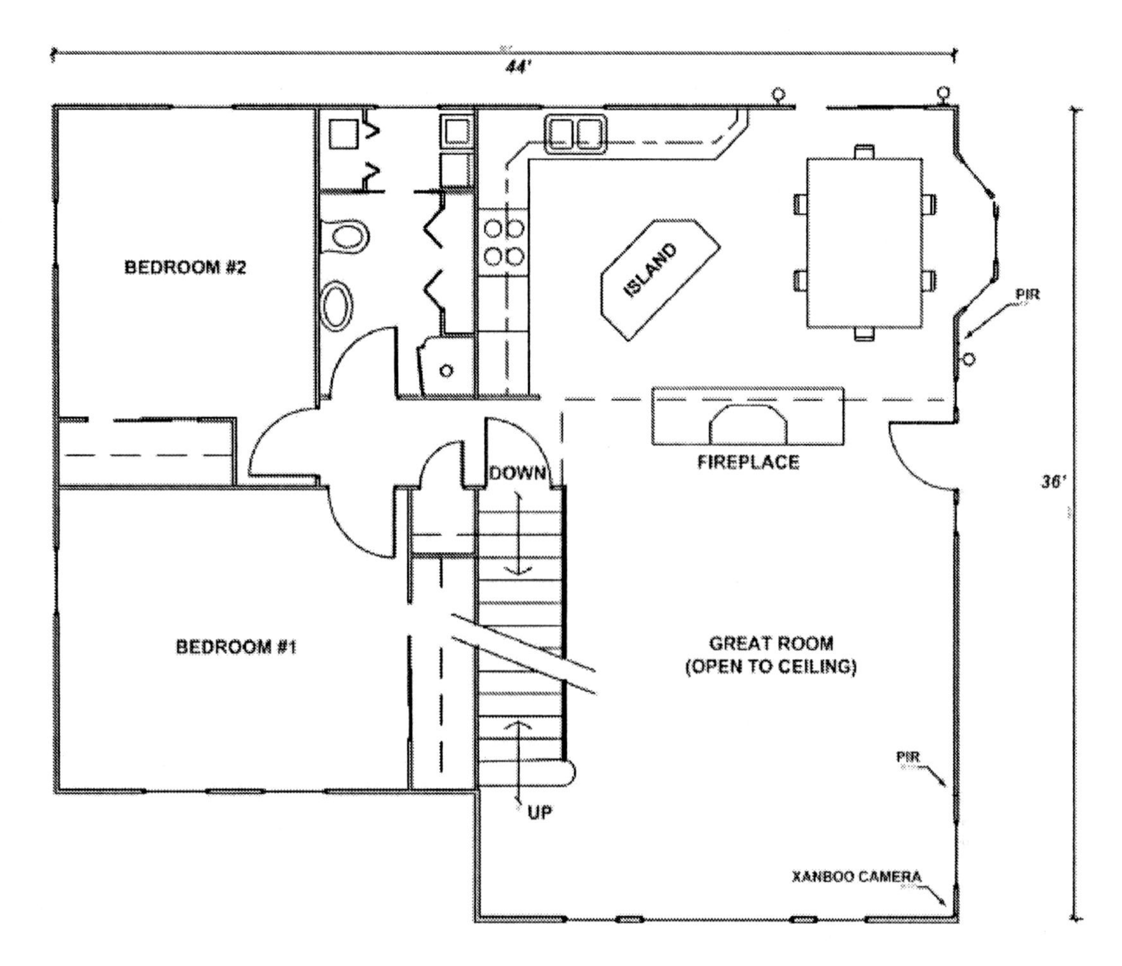

44'
36'
BEDROOM #2
ISLAND
PIR
FIREPLACE
DOWN
BEDROOM #1
GREAT ROOM
(OPEN TO CEILING)
PIR
UP
XANBOO CAMERA

1st floor

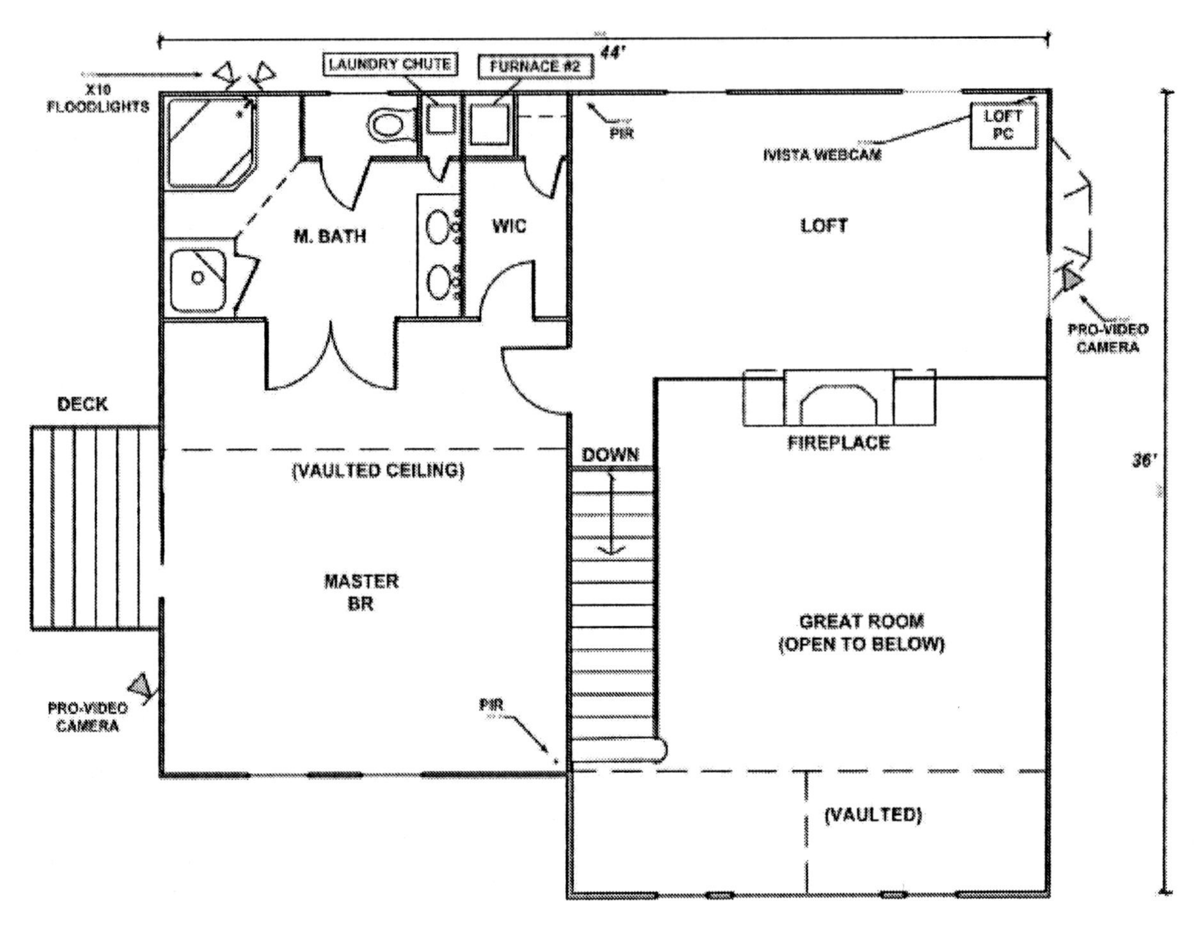

second floor

What I want to accomplish here is a sort-of three-fold objective:

- *Security*
- *Entertainment*
- *Practicality*

Knowing exactly what I wanted to do ahead of time for *security* was pretty easy: install a security system covering doors, windows, & interior movement.

The *entertainment* and *practical* aspects of it all will develop over the course of writing this book (and probably beyond). You'll get to hear me grumble & scratch my head as I troubleshoot - but I promise I won't throw anything and the language will be clean.

To begin with, I chose a central location for locating all my hardware (or at least as much as was practical). For me, this happened to be in an unfinished area of the basement where I could also locate my home office. I can leave clutter all over my desk, pieces of hardware & tools just laying around, & it's OK with my wife!

This (for all you thoughtful guys) is called planning ahead. By the way, you'll probably need your family's cooperation and patience as you implement some of your great ideas in home automation. So *do* try to keep them happy?

I ran wiring for the following, and ***it all terminates in that same "equipment room"***:

1. **Security:** 22 guage 2-conductor wiring to each door and window. I ran separate runs for each, though you could cover a number of windows in one room with a single run (simply looping the wiring in series from window to window). It was more work for me to run them all individually, but this allows for more flexibility in the future. I'll be able to group them as I wish in series, or set them up in their own zones.

   I also ran a Cat5 run for each of three interior motion detectors (Great Room, Loft, and Master BR). You really will probably only need 4 of the 8 conductors, but later on you might thank yourself that you have more wiring in place than you thought you needed.

   I also pulled a 16 gauge 2-conductor speaker wire to the attic and stubbed it outside for the siren. It will face my brother's house (who better to notify in a false alar- ... uh - I mean emergency?).

   What I didn't do was run anything for the keypad, so I had to retro-fit that after moving in. This is called *not* thinking ahead....

2. **Audio:** 16 gauge stranded (and twisted) 4-conductors individually run to volume control locations in six rooms: Great Rm., Loft, Master BR, Kitchen, 2 first-floor Bedrooms, and Back Patio.

   Hmmm ... OK, seven rooms.

   Anyway, along with the 4-conductors I also pulled a Cat5 for hard-wired IR sensors. This will not only provide input to the house audio, but will also function as another interface to Stargate.

   From the volume control/IR sensor location, two 2-conductors proceed to the appropriate in-wall/ceiling speaker locations. I drew up a "map" for the speaker wiring so that I could later (after drywall) cut in the speakers. Volume One has a discussion on speaker placement.

3. **Telephone:** Once again, individual Cat5 runs to all phone jack locations. It's important to "home-run" everything rather than loop from jack to jack, especially if you might install some sort of telephone system.

   You can certainly use less expensive wire for phones, but I wouldn't recommend it. Besides, this way you can always convert an unused phone jack into a data port (or the other way around).

   I also pulled enough Cat5 to accommodate all phone lines from the outside "SNI" (demarcation point) into my equipment room.

4. **Data:** The minimum standard you'd want to use is Cat5, though you can upgrade it to Cat5e or even Cat6. I pulled drops to the Master BR, two to the kitchen (one in the island & one in the wall), Loft, Great Room, & each of the other bedrooms.

   This will enable me to have a suitable network of multiple machines along with DSL access for all PC's.

   I talked in Volume One about taking care in running low-voltage wire. Not to repeat myself too much, but make sure you stay away from long close parallel runs with electric, & don't use the electrician's holes that he drilled through studs to run your wire. Drill your own. Probably the worst gremlin you can have is EMF, so do it right the first time when you run all your low-voltage wire.

5. **RS485 Run:** This is kind of a specialty item for Stargate, but it's really nothing more than another Cat5 cable. It allows the addition of up to sixteen RCS LCD96M keypads to Stargate, and also provides for an interface to RCS Smart thermostats. The nice thing about this is that it's just a single cable that daisy-chains from location to location.

   I pulled it from the equipment room to each of two furnaces (popping out on the wall nearby), and also passed through a couple of keypad locations. When you do this, just be sure to leave adequate slack at each location so you can actually work with the cable.

6. **Video:**
    - **CATV/Satellite:** While you typically only need a single RG6 coaxial cable to each outlet, I ran two. This allows for such future possibilities as running two sources to each TV or, if I should happen to locate a VCR or satellite receiver at a particular television that I want to share with other TV's - a path *back to* the equipment room with my video.

      Also, to the attic I ran two RG6 cables and coiled them up there. These will be for my Dish Network Satellite system. You might want to pull a cable for an antenna, even if just for FM reception. Wish I had.

    - **Security Cams:** While there are a variety of cams available, you'll probably be OK if you run an RG59 (smaller diameter than RG6) and 2-conductor (for power) to each location. I stubbed these out near the front door and also on the back side of the house.

      I'll also be installing some **webcams** on the inside of my house. The Ivista webcams operate on a simple "straight-thru" Cat5 data connection, so that I can actually plug these into any of my existing data jacks. Since I'll have more jacks than I'll be using at any one time, I really don't even need to run anything for these!

      Xanboo makes an inexpensive webcam that comes with a premade 9 pin DIN cable (about 100'). I ran this to my great room, but unfortunately assumed at the time that my PC would be in the loft. I suppose if I put another computer up there someday I'll be able to use it....

7. **Miscellaneous:**
    - I installed an **outdoor motion sensor (PIR)** at my front door, but didn't want to hook it up to my security system. That would have been a waste of a zone since I'm not trying to secure the great outdoors. I just want to know when someone approaches my front door. So instead, I ran this wire through the digital inputs on my Stargate and used the Caddx panel to power the PIR. Either way, the prewire was essentially the same: PIR to Equipment Room. I pulled a Cat5 (only needed 2 conductors) from the **doorbell** in the hallway to Stargate digital inputs with

the jumpers in the voltage position (the chapter on Security discusses this).

- Garage: Ran a 2 conductor from the **garage door** opener button (or you could run it to the opener motor itself - just so you can interface with the wiring) back to SG. This will later provide the capability to remotely open/close the garage door. I also made sure to install a security sensor on the garage door so I can monitor its up/down status.

- You'll need a Cat5 for your **RS232** (serial) connection from your PC to Stargate. In my case, the PC is located right next to SG so I really didn't need to run anything. If you have to run cable & build the ends, your computer store should have the RJ11 ends (6-pin kind, also called RJ12) & crimpers.

- For future microphones I pulled a 22-guage shielded 4-conductor to the kitchen, and also to the loft (volume control locations). If I should choose to add voice-activation to the system my wiring will be in place.

- If your PC isn't going to be located near all your equipment, it's probably a good idea to run a 22 or 24 gauge 2-conductor shielded (with ground) between the two locations. This will allow you to play .wav files on your system if/when you later want to do so.

- Lastly, I buried a 300' long magnetic automobile sensor alongside my driveway (I *do* have a long driveway). This went in at the same time as all my underground utilities.

8. **Theater:** This is just a little different because I didn't want to locate my components in the equipment room. I'll need them to be readily accessible & on-site. That made it necessary to run appropriate cabling from the cabinet that housed my A/V receiver & components back to the equipment room.

    - Coax from the Dish (passing through the equipment room) to my satellite receivers (I'm keeping them both in the "theater room").

    - Coax **back to** the equipment room for satellite distribution to other house TV's.

    - Coax feeding the modulated camera(s) (modulator in equipment room) to the VCR/TV.

- Several Cat5's for miscellaneous purposes: IR sensor located at TV, IR flashers to components, voltage-sensing of switched outlets in the receiver

- 4-conductor speaker wire to tie it all in to the house system.

When the prewire was done, I took notes on wire locations. That way I can seal up most of the wiring behind drywall where it can await the day that I'm ready to deal with it. I've sketched below a representative sample of my notes to give you an idea of how to do it.

Remember, you'll be doing this for YOURSELF to read. You don't need to be an architect. Just make it intelligible for YOU.

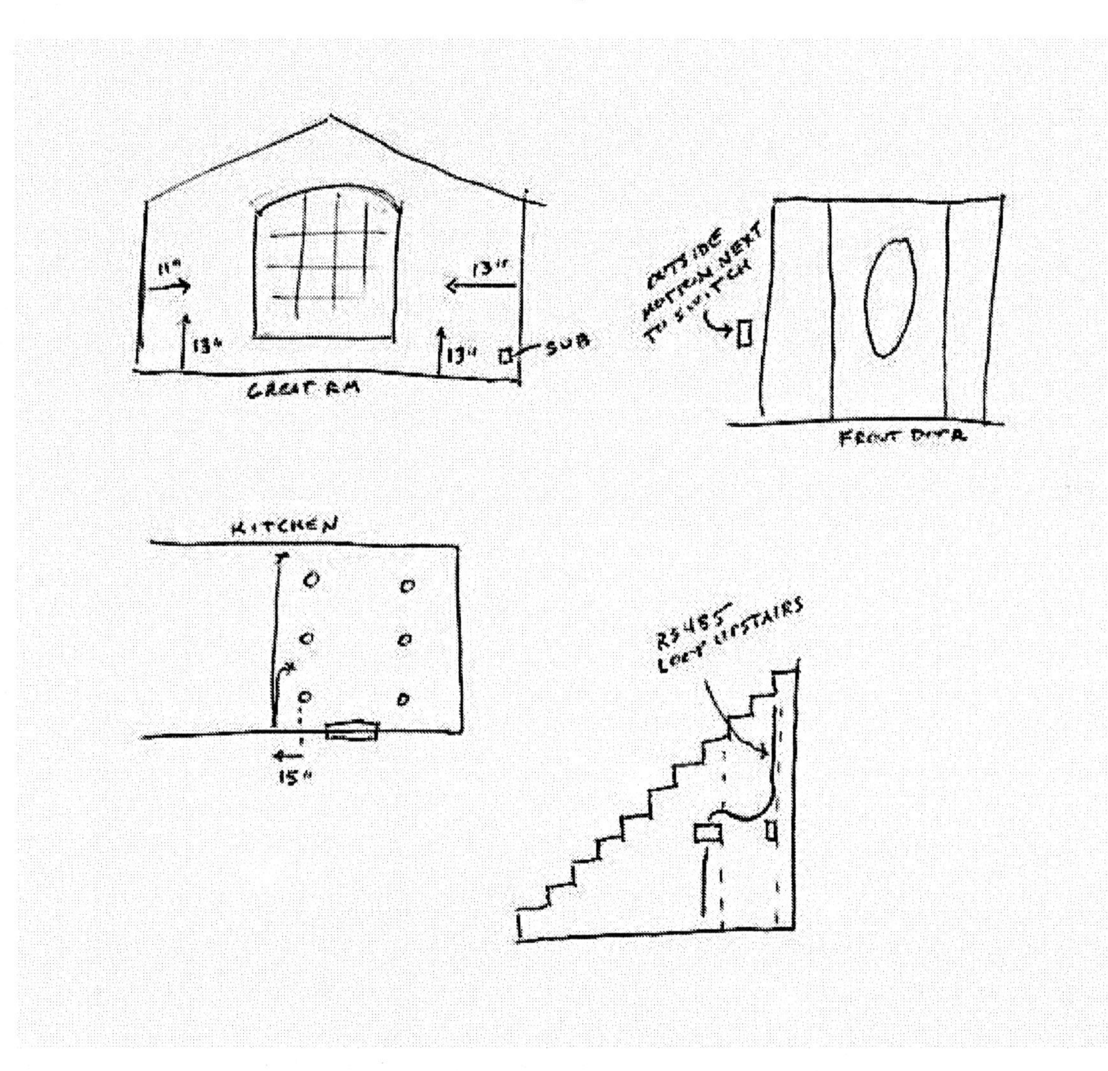

The Equipment Room:

Here's where all the fun happens. Truthfully, the toughest part about this is keeping it attractive. Below you see the 4x4 board I mounted for all my equipment, and the mass of wiring dangling beside it. Ignore the water tank - I have a cistern (*the price you pay around here for wanting to see the stars at night*).

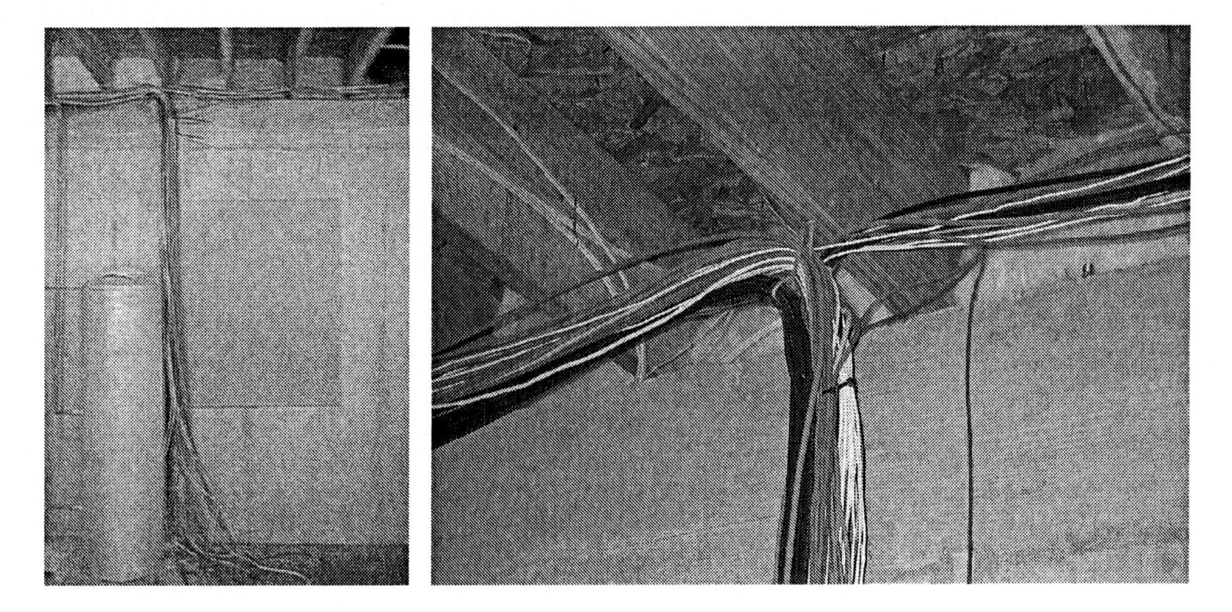

The picture on the right is a closeup of where the cabling falls from the ceiling. In order to keep it nice, I "split" the cables out near the ceiling according to their purpose, & tied them off with ty-wraps. Black is coax, white is security and data, blue is audio and miscellaneous cat5, the orange is coax for my cable modem.

It will look better when I'm done.... However, I want to stress that part of what I'm doing here is to show you how you can accomplish a lot without spending a lot.

There are some slick-looking structured wiring systems that I could have used - but I'm a bit of a pragmatist. If I can make it work without going nuts on expenses, I will. So yes, there are better looking ways of doing it, but to me that's not too important if it's in an unfinished basement.

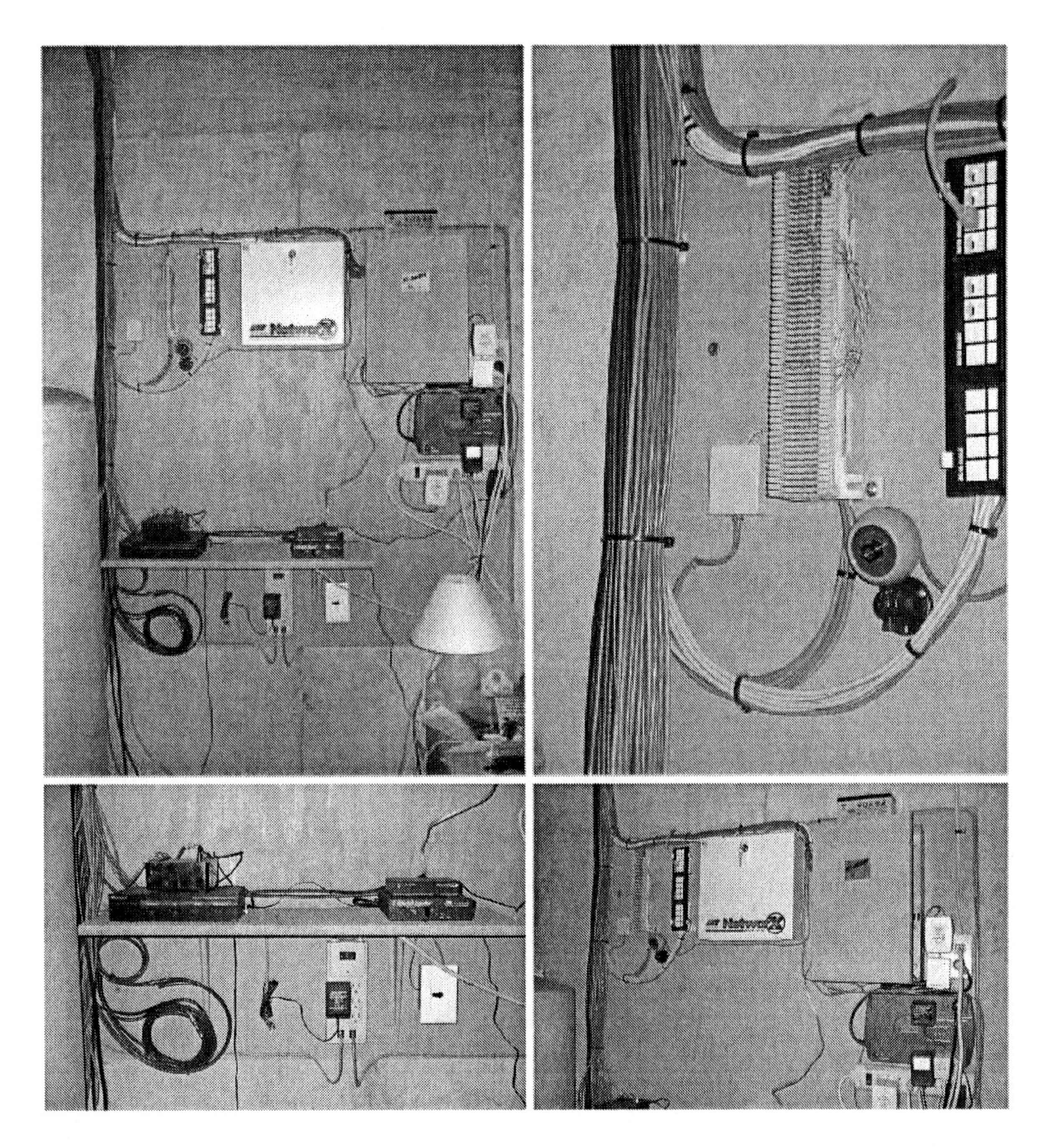

Essentially, I'm putting absolutely as much as is practical in this one place:

- Security panel
- Stargate
- Stargate IRXpander
- 66 block for telephone
- Patch panel for data
- PC
- Video modulator for camera(s)
- Speaker selector box for house audio

- Niles IRP-6 infrared junction box
- UPS battery backup

We'll go over each of these - how to connect them, how they interface, and what you can do once they're integrated.

... And to begin, let's start with something fairly simple: telephones....

Telephone Hookup:

At every home the phone company recognizes a "demarcation point", or "sni" as it's sometimes called. As far as they're concerned, this is where their responsibility ends and yours begins.

It's typically on the outside of the house (though it could be inside somewhere). It's generally also the place from which all your house wiring originates. However, you really don't want this to be the case. You'll want as much as possible to have everything "home-run" to your equipment room.

Therefore, I simply extended that demarc into the equipment room by running a Cat5 from the sni to my phone block (F.Y.I. a single 4-pair cat5 will accommodate up to 4 lines - one pair for each).

Looking at the pic below: I hardwired the cat5 FIRST OF ALL to a jack (*if you're particular you should look into using an RJ31 jack for your security system's sake*). This way I can bypass all my equipment if I have hardware problems. Plugged into the jack is the cabling which carries both of my two lines on a short little trip around the board before it even gets to the block.

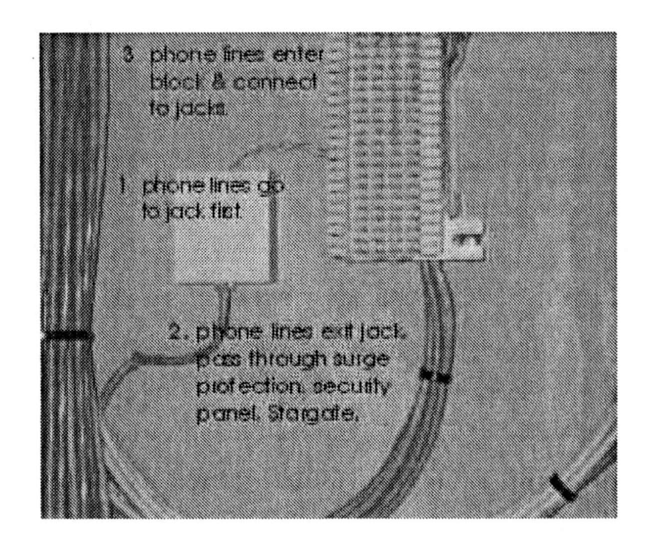

The first stop (below) is a surge protector for my phones. Power spikes don't just come in on the power line! Not to waste too much space with pictures at this point, I'll just say that from here line 1 passes through my Caddx NX8E security panel and Stargate, and then on finally to the block for distribution to all phone jacks.

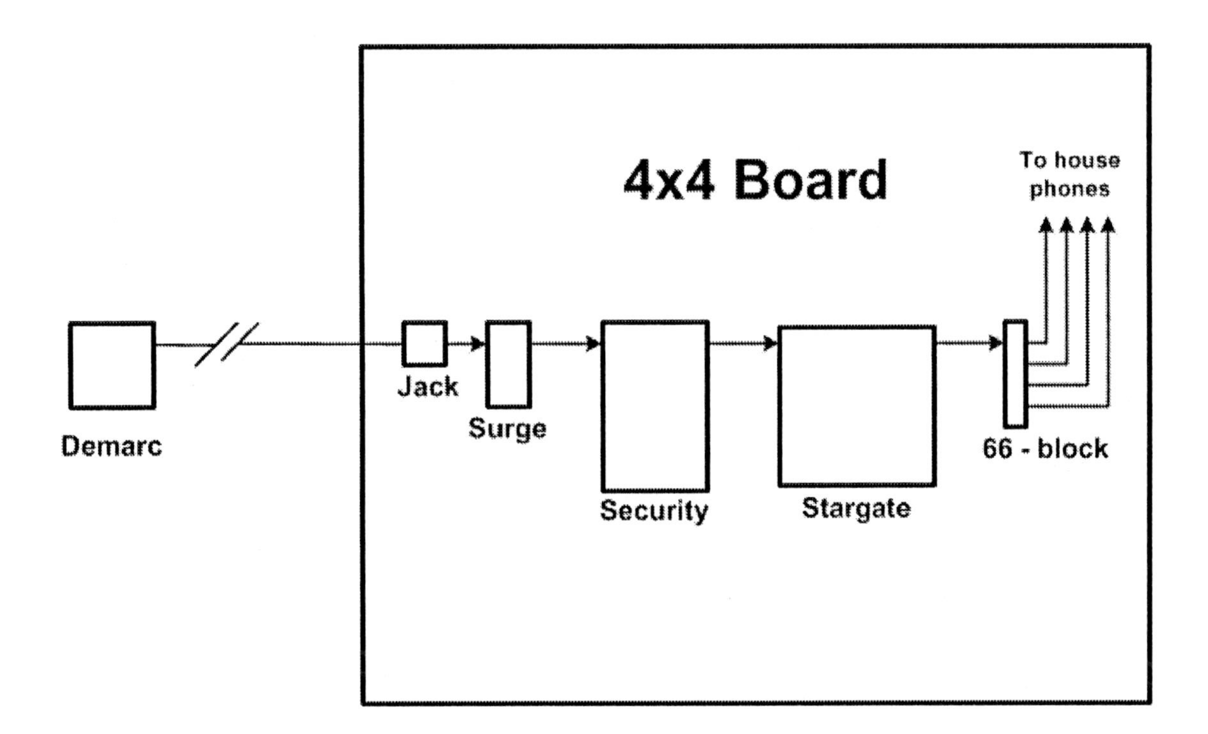

An important point to make: *if your security system is going to be monitored, make the security panel the first stop after surge protection. The security system needs to be able to seize the line in an emergency. An RJ31 jack will even interrupt existing calls to seize the phone line. Just make sure there are no other devices between it and the demarc.*

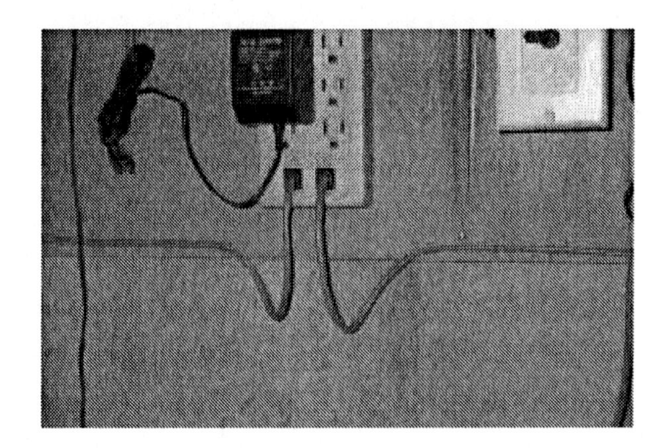

Before I get into punching wires down, let me explain a little bit about the 66 block. If you look at the next image, you'll see that this block has 4 vertical rows of pins. The pins that point in the same direction are already bridged together. This means that if you have a wire punched down in row #1 (far left vertical row), and another in row #2, they will have continuity (as long as it's on the same horizontal line of pins). The two rows to the right, however, are separate from the two rows on the left.

To make things neat (and to follow an industry standard way of doing things) I brought the incoming lines in on the far left side of the block, and punched the house jacks down on the far right side.

Since I have only ONE pair coming in for each line, but numerous pairs going out the house jacks, I needed some way to jumper them all together.

The pic below shows the jumpers I ran to bridge the two center rows of pins, thereby connecting the incoming lines to the multiple outgoing jacks. I highlighted the jumpers in white to make it easier to see what's going on here: the little "crook" at the top of the jumper connects to the pin in row #2 (remember that it's bridged to row #1?), effectively connecting the jumper to one of the wires in the incoming line 1 (in this case, blue).

I then continued the jumper down 10 of the pins in row #3 (which happens to be bridged to row #4). And guess what? Row #4 is where I punched down all my phone jacks (I can connect up to 10 this way). So now the incoming phone line (row #1) is connected to my jack (row #4).

I repeated this for each incoming wire (4 wires in total, 2 per phone line).

By the way, the colors that you see listed below are more or less standard colors for the wiring you'll find inside 4-pair cat5. Typically they'll be used in the following order: blue pair, orange pair, green pair, brown pair. Each pair consists of two wires (duh!), one of which is MOSTLY white with a little color, with the other being MOSTLY colored with little white stripes.

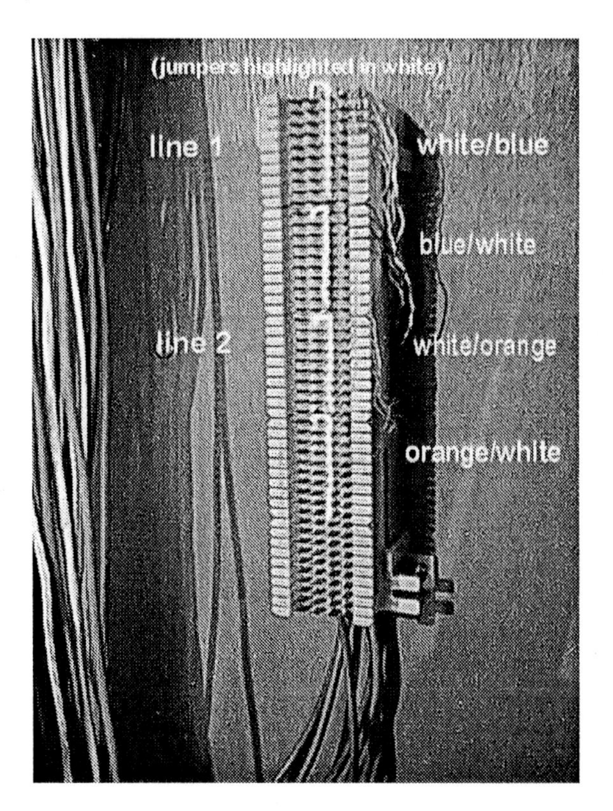

Now let me show you the tool you *ought to* use to punch these down. You can actually use a tiny screwdriver in a pinch, but you might end up with some shorts or bad connections.

You'll notice that beside the punch tool are two blades: a 66 blade and a 110 blade. The 66 (on the left) is what I used for the phone block. The 110 blade is what you'd use to punch down data jacks. While you might get away with a screwdriver on your phone block, I strongly suggest using the tool for your data connections. Heck, you might as well just buy the tool.

Another thing about each of the blades is the fact that one end of it has a cutting edge, and one end doesn't. When I punched my jumper down in a continuous fashion, I obviously didn't want to cut it until I got to the last pin. Otherwise, you'll usually use the cutting blade.

Now that I think of it, I used data (RJ45) jacks for my phone jacks. This works just fine. Though it might be a little more expensive, this will allow me to easily swap the function of the jack between data and phone in the future.

Most data jacks are color-coded, and you'll punch down on the blue color-coded pins for line 1. If you're using a two line phone, punch line 2 down on the orange pins (T568A).

However, if you're using a standard phone jack, line 1 is the red and green terminals; line two is yellow and black on the jack.

Oh, in case I didn't make it clear: if you only have one phone line, always use the line 1 terminals on the jack. A standard one-line phone always is wired up to look for line 1 (unless you get fancy with your wiring).

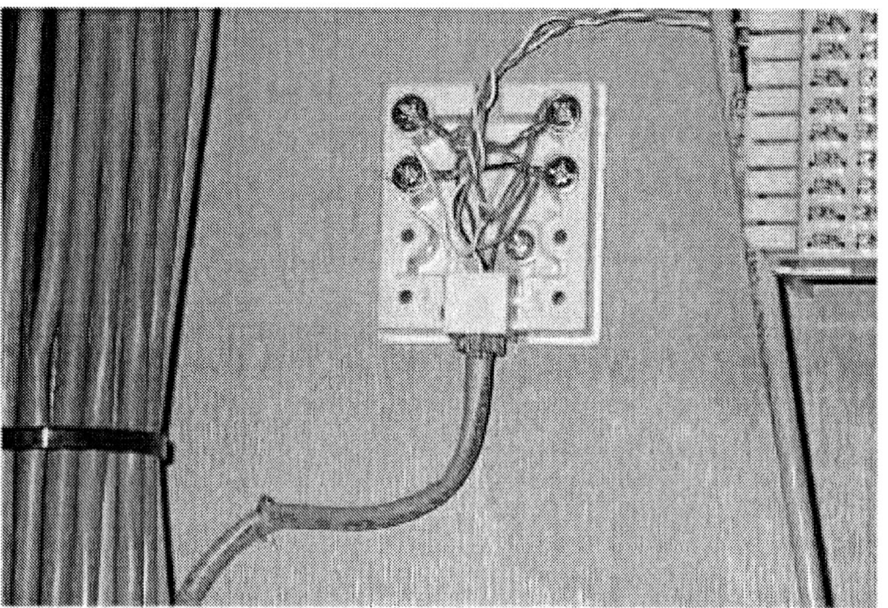

Standard phone jack

If you should need to put RJ11 connectors on your cords at some point, you'll obviously need a phone crimper. My suggestion would be to obtain a good crimper that does both RJ45 (data) and RJ11 (phone) connectors. It's possible to buy a cheap plastic phone crimper - but the last one I had (and the *only* plastic one) exploded in my hand when I squeezed it.

Okay, okay,... I should say it just broke. Anyway, I'll show more concerning terminating with crystal ends in the data section.

Here's a pic of the pinout on an RJ11. The most common RJ11 has only 4 pins, but some do have 6 as this one (sometimes called an *RJ12*). The important thing to see here is that the blue pair (line 1) is crimped on the two center pins. The orange pair (line 2) belongs to the pins on either side of the two center ones (pins 2 and 5).

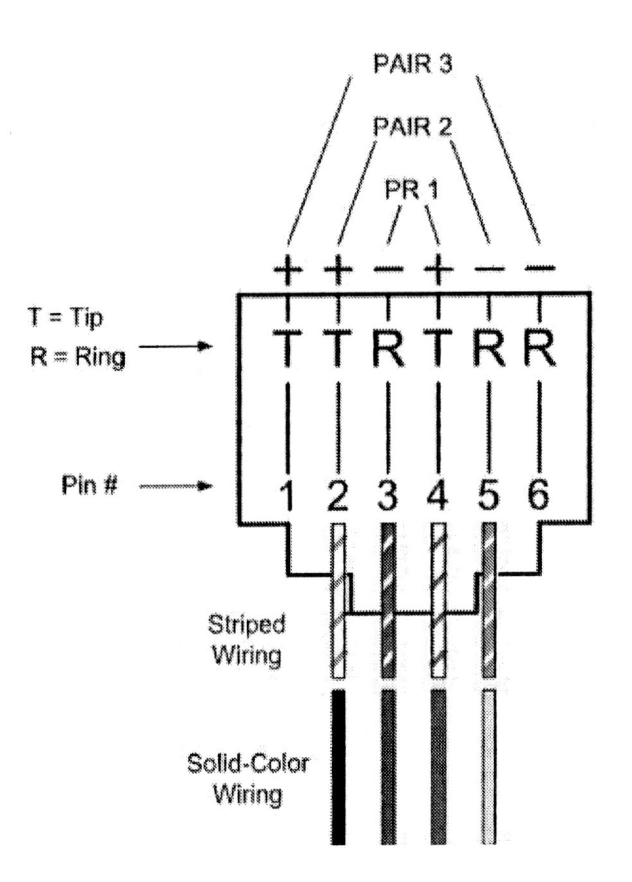

Some quick ideas of how to utilize the phone capabilities of Stargate include voice-mail, paging (via the intercom feature), caller ID announcements, & doorphones. Stargate's manual provides numerous sample scripts.

Data Network:

We're still on the peripherals of everything, even with your hi-tech new computer network. It's actually pretty simple, but in any decent smart home you should be able to communicate between PC's in different rooms. Even if you only have a single machine, it's nice to be able to plug it in virtually anywhere & still have access to your DSL/Cable modem.

If you only have dial-up right now and/or a single PC, maybe this isn't so important to you. But you know how things change....

The basic infrastructure wasn't a lot different than the phone cabling. I pulled all my cat5 from the equipment room to data jack locations, usually right along with the phone.

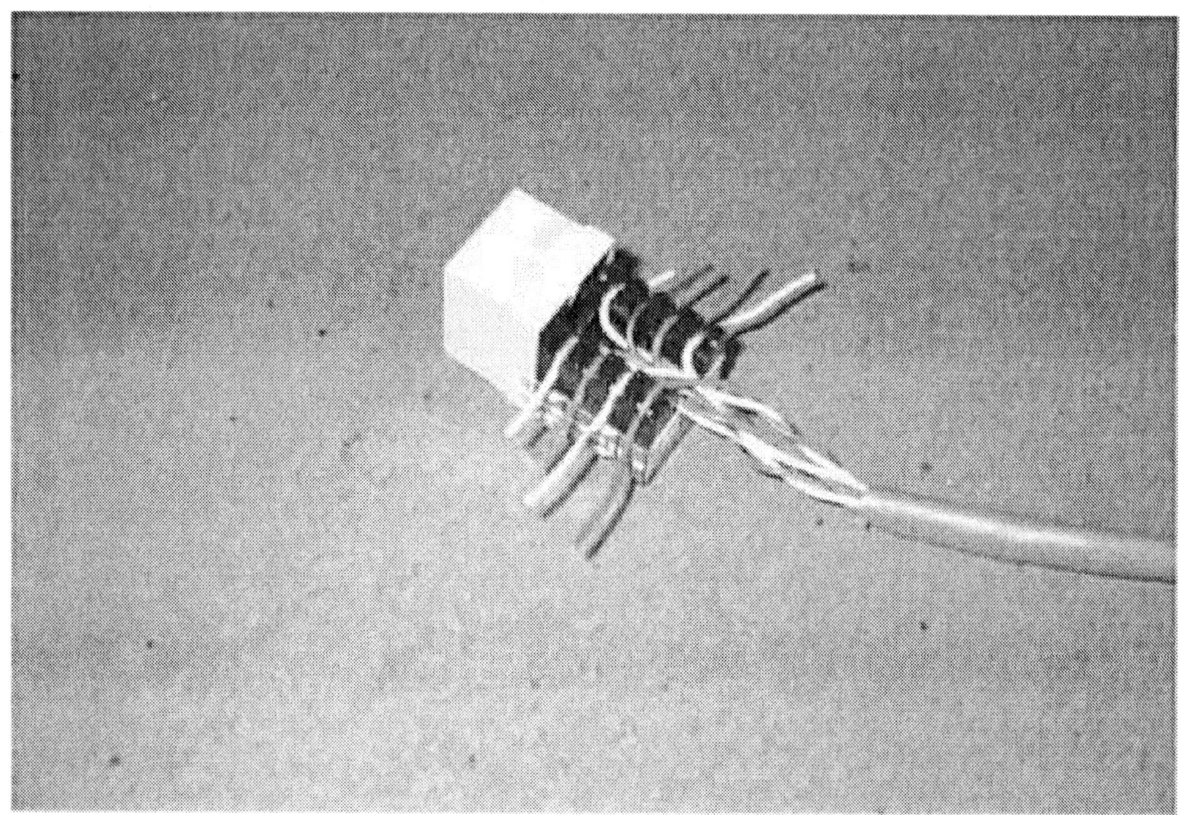

place the wires in the color-coded slots before punching them down. Leave as much twist in the wire as possible.

Terminating the jacks is obviously different than phone. I used Leviton jacks that are marked with T568A and T568B standards for the wiring pinout. If you look below at the two standards (below), you'll see that the only real difference between them is that the orange and green pairs are reversed.

It doesn't matter which one you use, but I used T568A. In either case the blue wires will be on the two center pins *(making it a perfectly functional phone jack if I choose to convert it to a telephone outlet)*. In that case, with the T568A standard pins 3 & 6 will also match up with the orange pairs for line 2.

Below is a (modified) copy of one of my newsletters that covered data networking:

Termination refers to punching your wiring down into jacks or patch panels, or just using RJ45 connectors (they look like big 8-pin telephone connectors). You can terminate your cabling in a couple of ways:

The simplest way (if you have the tools) is to just put RJ45 connectors on the ends of your cables. This eliminates the need for jacks or patch panels & will save you a few dollars. On the other hand, I think that wires sticking out of your wall (instead of from a jack in the wall) make for a not-so-clean installation - but if you do this you'll need to get your hands on a good set of crimpers. And if you really want a clean installation and you're installing a lot of drops, you might want to consider actually using a ***patch panel.***

It's better if you use ***jacks*** at all your PC locations. The jacks are all color-coded to make it a little easier. After all, the big mystery for most people is the order of the wiring pin-outs (what order to punch the wiring down).

Speaking of pin-outs, below are the standards you should use (I'll explain in a moment). As you look this over, bear in mind that white/green means a wire that's mostly white with green markings; green/white = green wire with white markings, etc....

TW568-A

1. pin #1 white/green
2. pin #2 green/white
3. pin #3 white/orange
4. pin #4 blue/white
5. pin #5 white/blue
6. pin #6 orange/white
7. pin #7 white/brown
8. pin #8 brown/white

TW568-B

1. pin #1 white/orange
2. pin #2 orange/white
3. pin #3 white/green
4. pin #4 blue/white
5. pin #5 white/blue
6. pin #6 green/white
7. pin #7 white/brown
8. pin #8 brown/white

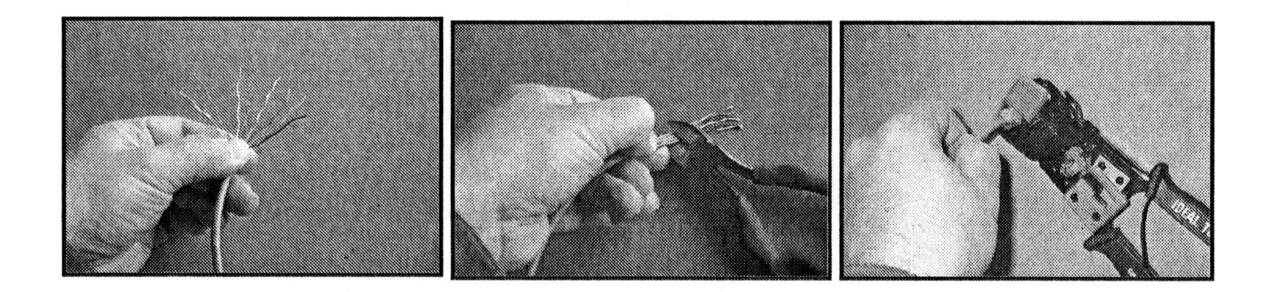

Begin by unwinding the wires with enough slack to give yourself some room to work.

After working them completely straight, cut them evenly to the proper length.

Slip the crystal RJ45 on. Make sure the wires go ALL the way in before crimping.

You don't even need to know the pinout if you're using jacks at both ends, but you will if you're using the RJ45's (F.Y.I. - Ethernet will actually only use pins 1,2,3, and 6 (green and orange pairs). The other wires are unused).

Unfortunately, things get a little more complicated at this point in the discussion. Most times you'll terminate your cable with the same standard (either one) at both ends - that's called a **"straight-thru"** connection - but occasionally you'll need to use 568-A at one end & 568-B at the other (a **"cross-over"** connection).

If you're connecting two PC's directly, you'll need a cross-over connection between them.

If you're connecting everything through a hub/switch, all your cabling will be straight-thru.

If you have a DSL Router, it will connect to your hub via cross-over & all your other cabling should be straight-thru. Also, some DSL routers have built-in hubs/switches. In this case, your PC's connect to it with a straight-thru cable.

Now, let's assume that you have an internet connection that you want to share with your entire network. If you're like most of us, your ISP is going to give you a single IP address - but that's only good for ONE computer!

Fortunately, there are several workarounds which can give all your networked machines access to the internet. Microsoft introduced in Windows 98 *"Internet Connection Sharing"* (not my favorite solution). Better though, if you have a DSL Router it *probably* has a feature called **"NAT"** (network address translation) built in to it.

NAT works by taking the IP address given you from your ISP, storing it in memory (sort of), and then allocating other private IP addresses to your machines. It keeps track of, and translates the IP addresses, ports, etc.

Gets kind of complicated...

However, the really important thing about NAT and your DSL Router is that is provides a reasonable degree of **"firewall"** protection. Especially important if you have an "always-on" connection, a firewall will help keep hackers out of your PC. It's incredible how many people there are who are cruising for hacking opportunities.... By the way, you can also get software-based firewalls at pretty reasonable prices.

Once you have everything up and running (network cards installed successfully & connected), you're ready to share files & folders. How you do that depends on your operating system, but I'm going to assume Windows 95/98 for the moment.

Go to "Start|Settings|Control Panel" and double-click the "Network" icon. Make your primary network logon "Client for Microsoft Networks" in the dropdown window. If it isn't there, click "Add|Client|Microsoft| Client for Microsoft Networks." The other thing you'll want to do is click "File and Print Sharing" and check the boxes. Restart your computer if you need to do so.

Sharing folders on the network is easy. Right-click on "My Computer" on your desktop, select "explore," and navigate to the folder you want to share. Once you've found it, right-click on the folder and select "sharing." You now have a dialogue for setting up your network share as you see fit! You're done! Within minutes of doing this, your folder will appear under "Network Neighborhood" on other machines with the rights you assigned (read-only or full-access).

I hope you absorbed all that. All my data connections are "straight-thru" from jack to patch panel.

I have a confession to make. I skimped on my patch panel. I bought what was intended to be a rack-mount 12-port patch panel & mounted it to a 66-block standoff. I had to butcher the standoff with a pair of sidecutters to make it work, and right now (unfortunately) the whole thing is held together with plastic ties.

I'm not telling you this because I'm proud of being cheap *(but I did save a few bucks).*

I'm telling you this because I have to show you a picture now:

By the way, at the end of this chapter is a "Networking Tutorial" if you'd like to know more of the intricacies of the subject. This is actually information that's published on my website (www.Integratorpro.com), but for your convenience it's also here in the book.

Once all this was in place and terminated, it was a simple matter of connecting all jacks to my cable modem. Since the modem has a single Ethernet port I had a choice of using a hub or DSL router to integrate the rest of the network.

Hubs don't generally offer much in the way of firewall protection, so I use a Linksys DSL router. Linksys, SMC, Watchguard, etc.... they all work pretty much the same: I have a cross-over connection from my cable modem to the Linksys, and straight-thru connections from all PC's (at patch panel) to the built-in switch on the Linksys.

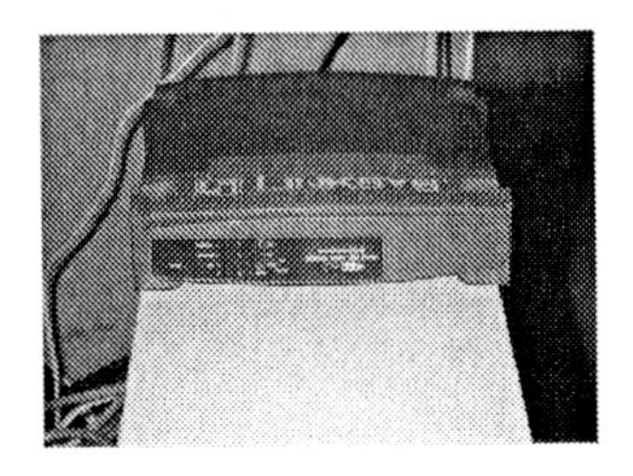

Note: Your ISP might provide you with either a Layer 2 OR Layer 3 modem/router. This just means that I don't know whether you'll need a cross-over or straight-thru connection from your ISP's device to your DSL router. If one doesn't work, try the other. There should generally be a "link light" on the device that tells you whether you have a good connection.

How to do your Own Home PC Network!

There are some **basic** things you need to know to get your home wired correctly for a PC Network. Beyond those few simple things, there are other issues like **pinouts** (the order of the wires) and **termination** (how to do jacks or plugs).

First of all, don't let yourself be too concerned about the variety of network topologies available to you. There's really only one standard 99.9% of us would use: **ETHERNET!**

That's a good thing, because Ethernet *(also called 10/100 base-T)* is *simple* in its design and installation. In fact, if you're only networking two computers, all you need to pull is a single Category 5 cable between the two machines. If later you decide to add other machines just pick one of the existing PC's as a hub location (you'll need a hub/switch or router), and pull cable to your new machines from there.

However, if you're going to do it *"right"* you should plan ahead for your home run location (where the hub or router goes). Pick a central location which offers the greatest flexibility in terms of getting wire runs to your possible PC locations.

But forsooth! *(think it's appropriate to say that)* ... I may be getting ahead of myself. After all, you may have a lot of questions about the exact nature of a "hub" or "router," or what in the world is Category 5 - and what the heck is a "pinout??"

So it might be a good idea to give some brief definitions.

TERMINOLOGY:

Category 5 - The label "Category 5" is simply a standard. Cat5 is very similar to telephone wire. In fact, a lot of telephone wire ***is*** Cat5. It's also sometimes called **UTP** *(Unshielded Twisted Pair).*

Inside a vinyl (PVC) jacket you'll find 4 pairs of twisted wires. The greater the twist, the more resistant the cable will be to RF & EMI (electrical noise) interference.

As a matter of record, there happens to be a Cat1, Cat2, and Cat3 standard as well. As the standards developed from 1 to 2 to ...5 the twist became progressively tighter (and the cable more expensive). *You can actually get Cat6, but probably won't need it in most cases.*

Hubs - Hubs are just simple devices for interconnecting (networking) computers. For the average home network, you can probably get a 5-port hub for about $40. One of the characteristics of a hub is that it passes all broadcast traffic to every computer on the network. Not a big deal when you only have a few machines, but it can cause real issues in bigger networks.

Switches - Switches are like hubs (they're both **layer 2** devices), but with a little more intelligence. They actually learn the paths from one machine to another (i.e. one port to another port), and then limit the associated broadcast traffic to the correct port(s). In larger networks, or networks with a lot of traffic, this means improved performance.

Routers - In order to understand all the advantages of using routers in a network, you'd need to understand the characteristics of **layer 3** (routers) versus *layer 2* (hubs/switches) devices. To put it simply, a router can "route" between networks using the **IP addresses**. However, routers popularly used on home networks typically perform a more distinctive function ...

Usually, "Linksys" or "SMC" (both just brand names) DSL Routers are used for their **firewall** features. These inexpensive boxes do in fact "route" by using a security feature called **NAT** (network address translation). The more expensive equipment by Cisco,etc. will do a lot more. In fact, more than you'll probably ever need.

Wiring Pinouts -
This just refers to the order that the eight wires are in when they're either crimped in an RJ45 connector or punched down in a jack. If you want to avoid making your own cables you can certainly buy them pre-made. However, if you're running cabling of significant length, you'll probably need to do it yourself.

Be aware that there are two different kinds: straight-thru and cross-over. When you understand the differences between layer 2 and layer 3 devices, it's easy to know which kind to use.

- Routers and PC's are layer 3
- Hubs and switches are layer 2

The rule is this: When connecting two devices of the same type, use a cross-over. When connecting two differing devices, you'll use a straight thru.

(Note: a funky twist on this is that most of the home-based DSL routers have a layer 3 WAN port, but the built-in port(s) for computers are layer 2 (hub or switch). In any event, you can look for "link lights" on the device to make sure you have a proper connection.

Network Interface Cards (NICS) - These are the little cards that you insert in your PCI or ISA slot that provide networking capabilities. The Ethernet cable plugs directly into it. Installing a nic is usually a pretty simple thing to do *(especially with plug and play Win95/98/2000/XP)*. Put it in, start the computer & follow the directions.

Yes, of course there are a lot more things to know - but that should be enough to prepare you for **what's about to follow ...**

... a Word about How LAN Communications work ...

For those of you who are Mac users, ***I'm sorry.*** *Most of this applies to the Windows environment, though references to TCP/IP still apply to you. I just don't know enough about Macs to speak authoritatively.*

Netbeui:
I've already mentioned broadcasts, routing, IP addresses, etc., & now I'd like to tell you why it matters which communication protocol you use. The simplest one available to you is **Netbeui**. It's so easy to use that it's probably the best thing for *very* small networks; however, it's not the most efficient.

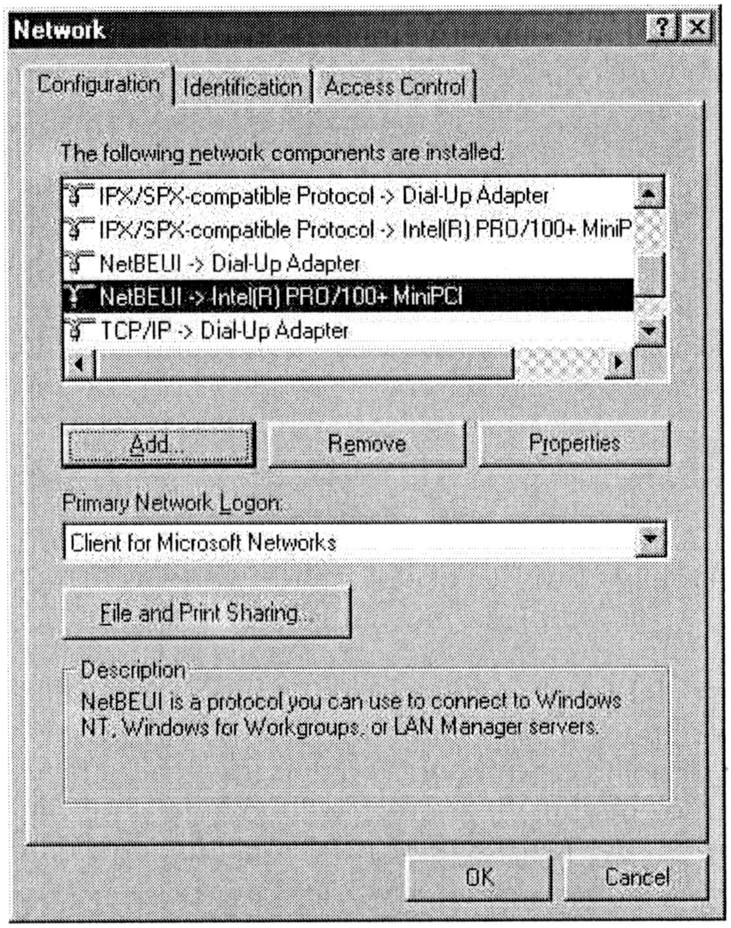

Network
Configuration | Identification | Access Control
The following network components are installed:
IPX/SPX-compatible Protocol -> Dial-Up Adapter
IPX/SPX-compatible Protocol -> Intel(R) PRO/100+ MiniP
NetBEUI -> Dial-Up Adapter
NetBEUI -> Intel(R) PRO/100+ MiniPCI
TCP/IP -> Dial-Up Adapter
Add... | Remove | Properties
Primary Network Logon:
Client for Microsoft Networks
File and Print Sharing...
Description
NetBEUI is a protocol you can use to connect to Windows NT, Windows for Workgroups, or LAN Manager servers.
OK | Cancel

(from Windows 98): You can install Netbeui (or any network protocol) by going to "Start, Settings, Control Panel." Then open the "Network" icon to see if it's already there. If not, click "Add, Protocol" (click Add again), select "Microsoft," then "Netbeui" in the right column, & click "OK."

Netbeui is a broadcast protocol, meaning that its message goes to every computer on the network. If you're looking to access a file on a particular remote PC (let's say from your laptop), all the communication between the two machines is heard as *noise* by the other machines. They have to listen even if they have nothing to do with what's going on.

In a bigger network with multiple conversations going on at the same time this can be a problem. ***Why?*** Because on an Ethernet network, computers have to take turns sending data. They're continually *interrupting* each other, which can degrade network performance.

To sum it up, Netbeui is fine if you have two or three machines to network in your home - but if you have an Internet connection to share between them you'll need to use **TCP/IP**.

IPX/SPX

Not going to spend much time on this. If you need to know about it, you probably already do. It's generally used in Novell environments, & is just about as simple to install & use as Netbeui. It relies on hardware addresses burnt-in to the network cards to identify individual PC's.

TCP/IP

Whoa! Easy to write an entire book about this - or several, depending on how deep you want to go. This is the communications protocol of the internet, & actually of most Local Area Networks (LANs) today. If you're taking your network at all seriously, use it.

Again, it's installed the same way as the above Netbeui example, & whether you actually assign an address to your network card will depend on what you're doing & who you're doing it with. Usually, if you have a DSL/CABLE modem your IP address will be automatically assigned to you. If you have a "static" (unchanging) IP, you'll have to enter it manually.

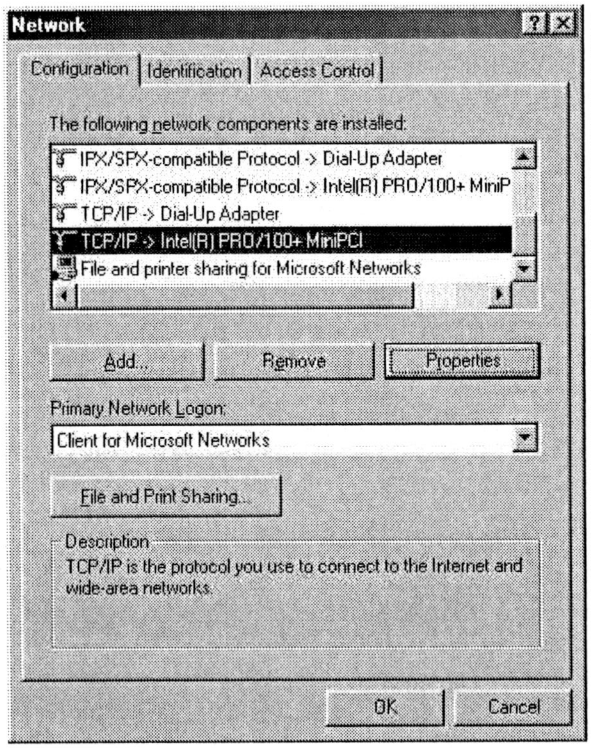

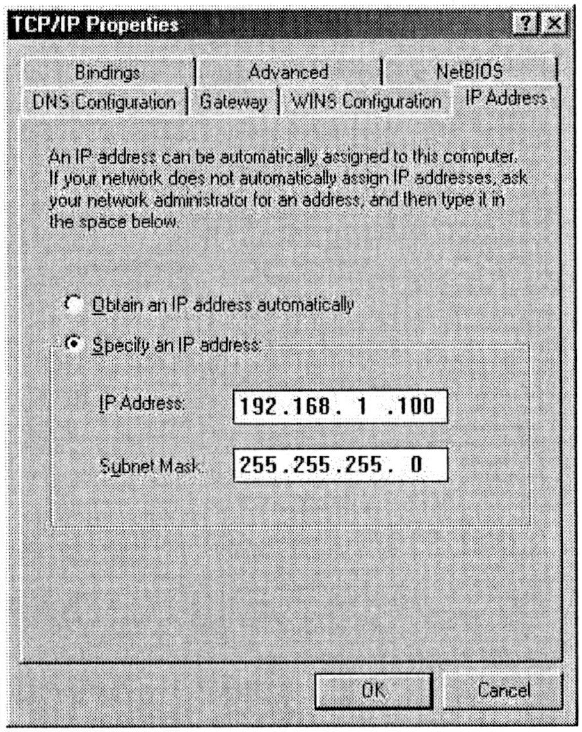

(from Windows 98): From Control Panel, select "Network." "Add" if you need to. If it's already there, highlight "TCP/IP-> (your network card)" & click "properties." Choose to either obtain an address automatically or else enter your own.

Following is a more in-depth discussion on IP addressing. If you're creating your own network scheme, you'll need to know about this.

My assumption, though, is that you'll only need to know a few things ...

IP Addressing:

It's a struggle to keep this simple, so I'll probably start out that way and then make it as complicated as it can be *(just kidding - I think...)*.

In order for PC's to communicate using IP addresses, they have to be on the same range of network addresses. A given range of addresses (called a network or subnetwork) is determined by a combination of two things:

- IP Address
- Subnet Mask

To start out, IP addresses consist of 4 groups of numbers separated by dots (e.g. 192.168.1.100). These are actually *decimal* representations of *binary* numbers. Each of these is referred to as an *OCTET*, because in binary form they each consist of eight digits.

Furthermore, part of this address represents a "network" address (which all machines on the given network will share in common), and the rest of it represents the unique PC address.

I know this is confusing if you don't know binary, but bear with me - I'm just trying to communicate the geek language to you so things will make sense as we go along...

BINARY

You're probably well aware that computers use a digital language, and generally that language is binary. In binary, you only have 2 numbers. Actually, you don't really have numbers at all, though they're represented by 1 (one) and 0 (zero). It will make more sense for you to think of it as two states (i.e. *on/off* or even *checked/unchecked*).

When you count in binary, you count exponentially. If you were to list a group of 1's like this ***"11111111"*** each digit would have the following decimal values (counting from **right to left** no less!):
"1,2,4,8,16,32,64,128."
Once you've digested this, read on ...

You would then add all the values together to get the decimal (NORMAL/UNDERSTANDABLE/HUMAN LANGUAGE) equivalent: ***255.***

If, on the other hand, one of these binary "1's" was a zero, you'd simply subtract it's value from the total as in the below examples:

- binary "11111110" = decimal "128,64,32,16,8,4,2,0" for a decimal total of **254**.
- binary "11111101" = decimal "128,64,32,16,8,4,0,1" for a decimal total of **253**
- binary "11111011" = decimal "128,64,32,16,8,0,2,1" for a decimal total of **251**
- binary "10111101" = decimal "128,0,32,16,8,4,0,1" for a decimal total of **189**
- binary "00111101" = decimal "0,0,32,16,8,4,0,1" for a decimal total of **61**

You might have to study that for a minute to grasp it. It's important to understand this ***if*** you need to manage your network schemes. Mostly, it comes into play when you combine it with your subnet mask.

However, you might not need all this. In fact, most people probably don't. I said earlier that part of the address represented the network, and part of it was the unique PC address. The subnet mask tells you which is which. Consider the below address and mask:

ADDRESS:
192.168.1. *28*

SUBNET MASK:
255.255.255. *0*

If you were to line up the bits on these one above the other, you'd have:

11000000.10101000.00000001. *00011100* *(IP address)*
11111111.11111111.11111111. *00000000* *(Subnet mask)*

Imagine that the subnet mask is printed on a transparency, and you can lay it directly on top of the IP address. All the "1's" of the subnet mask correspond to (or "mask out") the network address *(in **bold**)*. All the remaining bits of the IP address equate to the PC's unique ID *(in **italics**)*.

Let me simplify. There are three classes of IP address schemes that are common:

- Class A: has a subnet mask of 255.0.0.0
- Class B: has a subnet mask of 255.255.0.0
- Class C: has a subnet mask of 255.255.255.0

If you aren't concerned with internet access, you can configure your network anyway you want using one of these classes. But my guess is that you probably ***are*** interested in accessing the Internet. So here's how you deal with your Internet Service Provider (ISP) ...

Dealing with your ISP:

Most of us are allocated a single *dynamic* IP address by our ISP. This means that it's subject to change occasionally. It also means that you can only connect ONE PC to the Internet at a time.

To address this, Microsoft has introduced ***Internet Connection Sharing*** in Windows 98 & above to allow multiple PC's to share a single internet connection. You can find out how to do this in the "help" menu, but I'm personally not too fond of this solution (I think it has "issues").

A better way is to purchase a SOHO (small office/ home office) **broadband router** (see the Linksys or SMC DSL Routers). You'll usually find that they have a built-in switch or hub for several PC's as well. If you're using **dial-up** you can do this too, though you should make sure your desired router has this capability before you buy it.

Linksys BEFSR41 *SMC Barricade* *Netgear RM356*

There are several nice things about these devices. One, they perform **NAT** (Network Address Translation), and in the process provide a good degree of ***firewall*** protection ...

And now, I must digress again. It's time to discuss firewalls.

You really should exercise a degree of caution over internet security. Most hackers are probably benign & just like to see what they can do. However, I know I don't want them practicing on ***my*** machine & I suspect you don't either.

When a broadband router performs **NAT,** it takes the "public" IP address that your ISP gives you, and ***translates*** it into a "private" IP address. The router then gives this new address to your computer. When you initiate a session, the router does the reverse. It translates your "private" address into the "public" one. There's a literal and very real distinction between public and private IP addresses.

When standards were developed for IP, it was determined that certain address ranges would be designated for private use, and would be "unroutable" (useless) over the Internet. ***There are three address***

ranges designated as private:

10.x.x.x *(Class A networks that start with 10.)*

172.16.x.x *(Class B networks that start with 172.16.)*

192.168.1.x *(Class C networks that start with 192.168.1.)*

With few exceptions, everything else falls into the **"public"** (routable) domain. When the typical hacker gets to your firewall, unless he's very good he'll probably not be able to get into your network..

The other thing that NAT does is to make it possible for multiple machines to use the same public IP address at the same time. Now, if you're thinking, you might wonder **why replies from a web site don't go to *all* the computers on the network.** After all, they share the same public IP address.

The answer is that while NAT translates all the different "inside private" addresses into a single "public" address, it uses different **"ports"** for each computer. So all the machines identify themselves to the internet with the *same IP address,* but with *different ports.* *(F.Y.I.: The combination of IP address and port is known as a **socket**).*

This way you can have multiple simultaneous internet connections from your network without confusion.

The Breaker Panel:

Here's where I could get into trouble. I've actually gotten flamed from a few people in regard to electrical code, etc., so let me say upfront:

Double-check with an electrician about code in your area regarding how you connect the devices I'm going to discuss.

That is, *IF* you're doing this yourself. It's not real time-consuming, so if you want to hire an electrician to do it, it shouldn't break you anyway. As a disclaimer, let me point out that Leviton says that the installation "must" be done by a qualified electrician.

It's a good idea to install a couple of things here. A **whole-house surge suppressor** will add a degree of protection to your electronics (especially X10 switches, etc.). Secondly, a **signal bridge/repeater** can aid tremendously in your X10 communications.

Actually, these are both optional. The suppressor is a plain old good idea. The bridge/repeater *might* not be necessary. It's something you could come back and add later if you experience X10 problems.

Whole House Surge Suppressor: I used a Leviton unit which mounts externally to the breaker panel. While I had to insert a small length of electrical conduit between the Leviton device and a knock-out on the breaker, it was relatively simple.

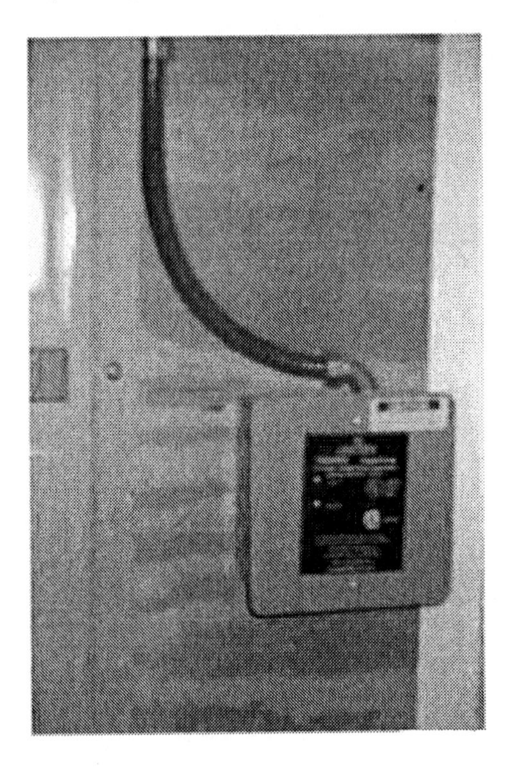

Especially since I chose to be cautious and use my electrician-brother to make the connections. Actually, if you know how NOT to electrocute yourself you can do this. You know, wear rubber-soled shoes, throw the main breaker (off), don't touch the wrong stuff, etc. The wiring diagram that comes with it is relatively self-explanatory.

SPD Performance Specifications—Cat. No. 51120-1, -3	
Rated Single Pulse Transient Energy (10x1000 µs Joules)	950
Maximum Single Pulse Transient Energy (8x20 µs, Amperes Peak)	50,000
Response Time	Instantaneous
Cat B3 Combination Wave (8X20 µs)	480V
UL 1449 Rating (L-N)	500V
Physical Specifications	
Operating Temperature Range	-10 to 60°C
Storage Temperature	-20° to 65°C

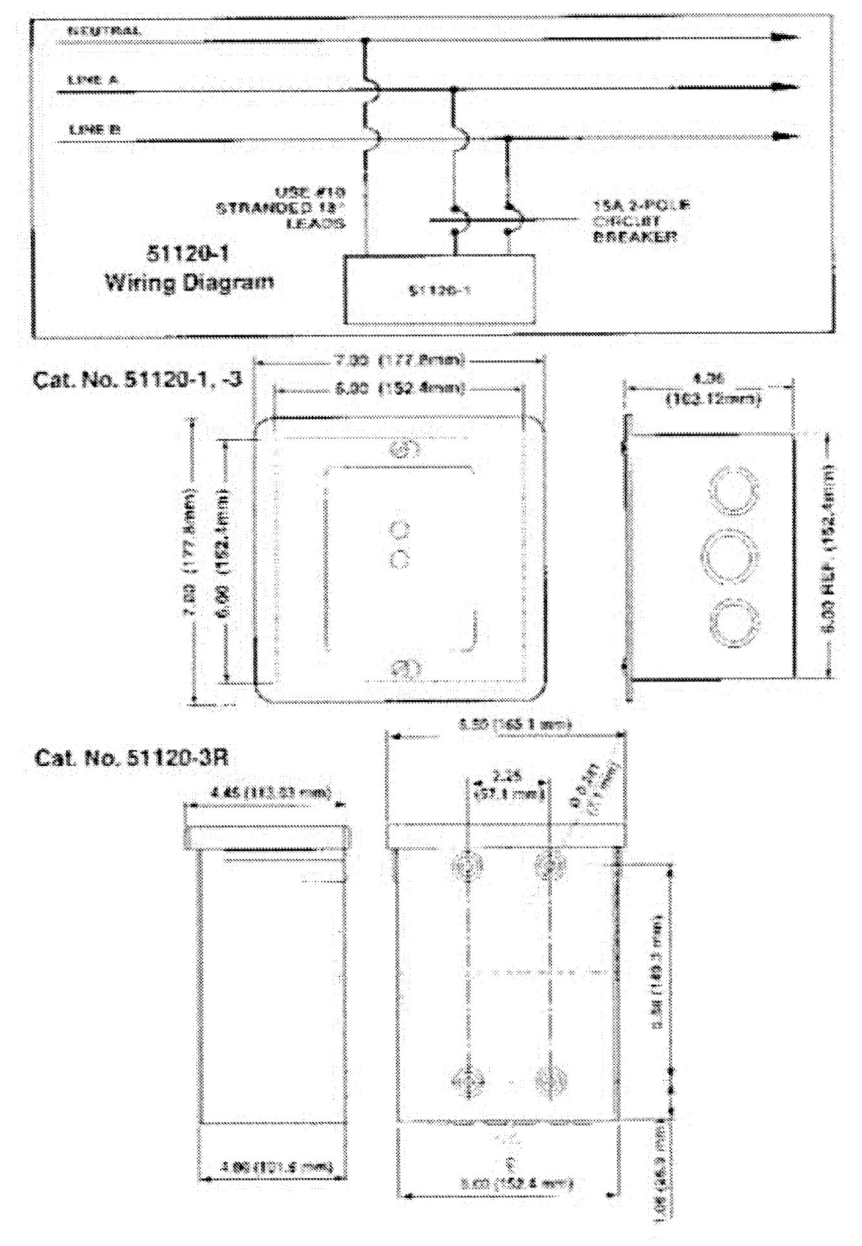

from Leviton instructions

As the top half of the diagram shows, one lead from the surge suppressor goes to the neutral bar, and the other two to a 15-amp breaker. Easier than Caddx! Of course, the security panel won't toast me if I do the wrong thing....

Signal Bridge: Here's where I've gotten a little negative feedback from some people concerning the way I connected mine.

I simply installed my bridge right inside the breaker box. It may not be code (so I'm not recommending you do it this way), but it works for me. However you do it, what really matters is that the red and black leads go to either side of a 15-amp circuit breaker; the white goes to the neutral bar. Not much different from the surge suppressor.

X10 signal bridge

X10:

X10 can be used for a tremendous number of things. You can use it for security, contact-closure, lighting - just about anything high or low voltage.

However, nice as it is, the less it's used, the better. Anything hard-wired is superior. Response time is faster, don't have to worry about signal collisions or other related issues....

I used X10 for lighting and for a couple of outdoor floodlight/motion detectors. That's about it. Actually, setting it up is so simple that I don't think I need to waste a lot of space telling you how to set addresses, etc. All that info will come with whatever modules/devices you get for yourself.

At the end of this page I do, however, include a link to a few other ideas for X10.

Lighting: One of the nice things about using X10 for your lighting is that it doesn't require any special preparation as far as wiring goes. As long as you have a neutral (and most newer electrical layouts do), you should be good.

One thing you want to be alert about is the total wattage. The typical X10 switch will handle up to **600 watts**, so if you add up the wattage on all light bulbs controlled by any particular switch (or group of switches), and it's more than what your switch says it can handle, look for something more heavy-duty.

Of course they cost more, but X10 switches are available that will do **1000 watts**. One other thing to be aware of is that, due to the heat generated, grouping any switches together will typically reduce the switches' wattage capabilities.

I chose to use some relatively good quality X10 switches: Switchlinc 2-ways. There are several brands of switches you can consider: Leviton, PCS, and Switchlinc. Of these, I'd personally recommend either PCS or Switchlinc. In fact, if you'd like the option of upgrading to a star/hub based configuration (for improved performance), you probably *should* consider PCS.

I happened to go with the Switchlinc brand simply as a matter of economics (got a good deal on them). At one point the manufacturers (Lightolier) had a serious issue with these. About 2% of their switches were behaving strangely, coming on/off on their own, etc., and a lot of people had complaints (including me).

Since then Smartlinc has assumed responsibility for manufacturing Switchlinc, and they've apparently dealt with the problems. I haven't experienced any of the old symptoms to date and everything seems to be fine. By the way, Lightolier still makes the old switches under the "Compose" brand name.

> Switchlinc also makes a "lite" version of their 2-way switches, the difference being that the 2-way also *sends* an X10 signal as well as *receiving* one. I don't personally find much of a need to send X10 signals from my light switches, but it's nice to know that I can if I need to do so.
>
> The drawback to the 2-way switches (aside from cost) is that they introduce a lot more X10 traffic to the powerline. It's recommended that you don't use more than 10 of these in your application.
>
> One other note: here "2-way" refers to the switch's ability to send/receive X10. More commonly the terms 2-way, 3-way, etc. refer to the number of switches controlling a common load.

LED indicates light level

If you have a simple load (single switch) the connections are extremely simple (see the diagram on the left below). You do, however, need to distinguish between the existing "line" and "load" wiring, both of which are likely to be black wires in your receptacle box.

The easiest way to tell is with a voltmeter, testing for the presence of power between a ground wire and black. "Line" should have power on it and "load" (which goes to the *incandescent* light fixture) should not.

Another way to tell is if you see a black wire looping from one switch to another. This is likely to be "line."

SwitchLinc Wiring Diagram

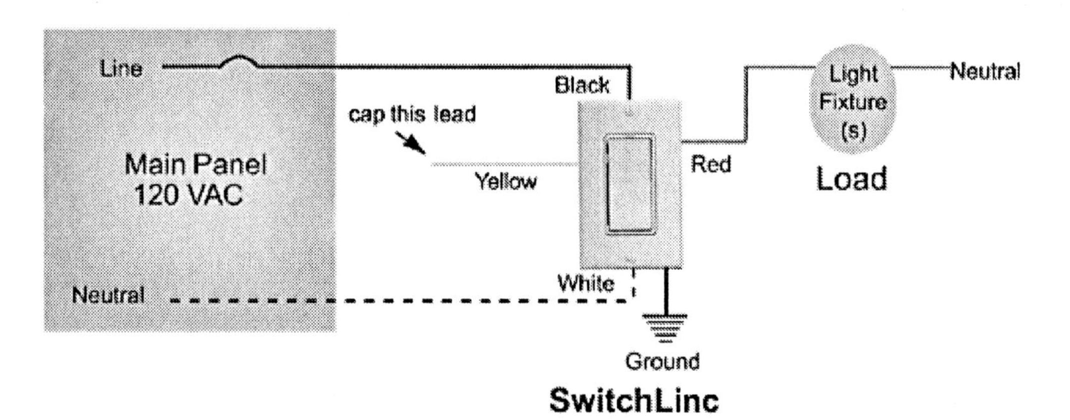

SmartLinc Multi-Way Wiring Diagram

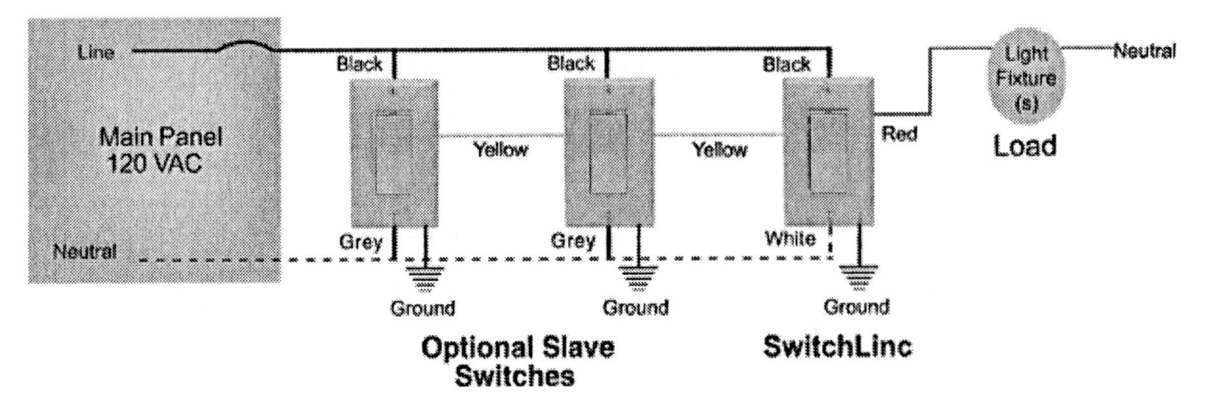

Using 2 SwitchLincs to Create a Virtual 3-Way Circuit

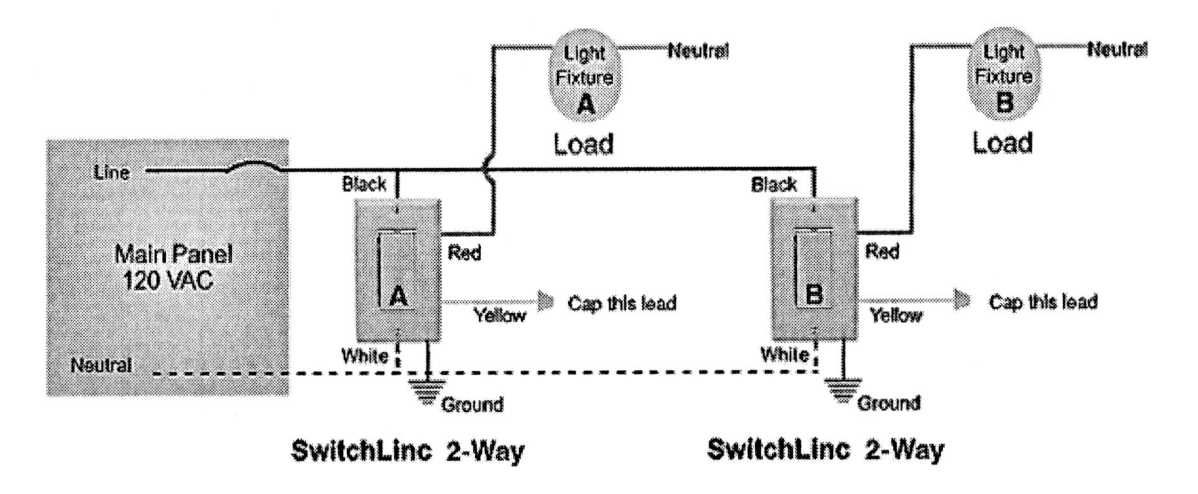

Light fixtures A & B will work as if they were part of the same 3-way circuit. Either switch A or B can be used to turn on/off (dim/brighten for dimmers) lights A & B even though there is no "traveller" wire.

As you can see from the diagrams, it gets just a little funkier when you have multiple switches on a single load, but you can do it! If you aren't confident, or if you weren't going to wear rubber-soled shoes & turn the power off beforehand if I hadn't told you to do so,

... you'd better get an electrician.

Setting the address on the Switchlinc switches is done electronically rather than physically adjusting a dial (like on an X10 module). By default, all the switches are set to A1.

Basically, you press a button on the switch to prepare it for programming. From another controller you send the X10 signal bearing the desired address for the switch, and when the switch detects it, it accepts the new address as its own.

Some other nice features are the ability to set the "ramp rate" (speed at which the light brightens/dims), preset dim level, and (somewhat more advanced) grouping lights together into "scenes." The instructions are easy enough to understand.

X10 Floods: Again, the only preparation necessary was for me to tell my electrician to wire for floodlights in the front and back of my house. These Leviton X10 floods are configurable for sensitivity to light and motion. Pretty simple to do, actually.

The practicality of it is that it's an easy way to get both lighting control AND motion detection on the outside of the home; and since they *send* X10 as well as *receive*, I can set my entire system up to respond however I choose when motion is picked up.

X10 Floodlights: set the send/receive address and sensitivity

---

See below for more about **X10 theory**. This is a reprint of an article found at SmarthomeUSA.com.

How X10 Works

X10 communicates between transmitters and receivers by sending and receiving signals over the power line wiring. These signals involve short RF bursts which represent digital information.

X10 transmissions are synchronized to the zero crossing point of the AC power line. The goal should be to transmit as close to the zero crossing point as possible, but certainly within 200 microseconds of the zero crossing point. The PL513 and TW523 provide a 60 Hz square wave with a maximum delay of 100 µsec from the zero crossing point of the AC power line. The maximum delay between signal envelope input and 120 kHz output bursts is 50 µsec. Therefore, it should be arranged that outputs to the PL513 and TW523 be within 50 µs of this 60 Hz zero crossing reference square wave. .

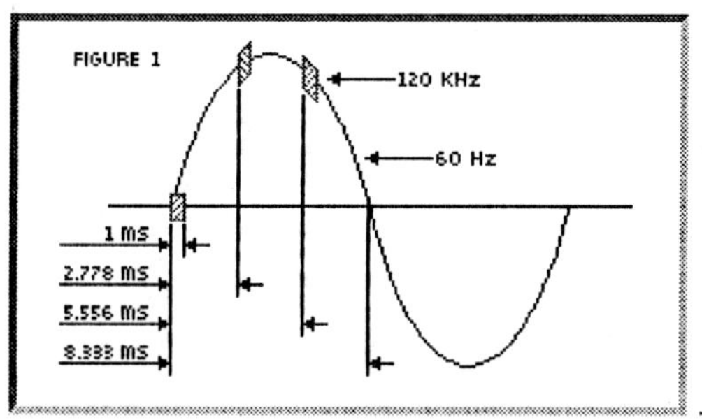

A Binary 1 is represented by a 1 millisecond burst of 120 kHz at the zero crossing point, and a Binary 0 by the absence of 120 kHz. The PL513 and TW523 modulate their inputs (from the O.E.M.) with 120 kHz, therefore only the 1 ms "envelope" need be applied to their inputs. These 1 millisecond bursts should equally be transmitted three times to coincide with the zero crossing point of all three phases in a three phase distribution system. Figure 1 shows the timing relationship of these bursts relative to zero crossing. .

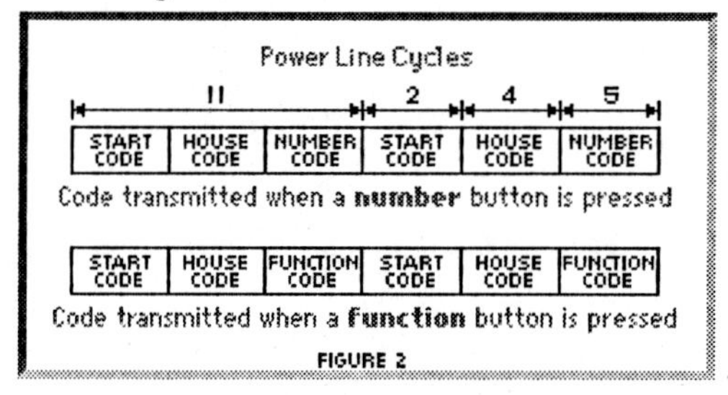

A complete code transmission encompasses eleven cycles of the power line. The first two cycles represent a Start Code. The next four cycles represent the House Code and the last five cycles represent either the Number Code (1 thru 16) or a Function Code (On, Off, etc.). This complete block, (Start Code, House Code, Key Code) should always be transmitted in groups of 2 with 3 power line cycles between each group of 2 codes. Bright and dim are exceptions to this rule and should be transmitted continuously (at least twice) with no gaps between codes. See Figure 2. .

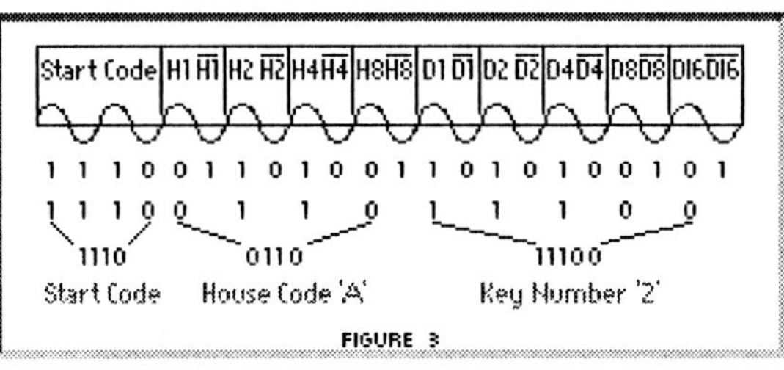

FIGURE 3

. Within each block of data, each four or five bit code should be transmitted in true compliment form on alternate half cycles of the power line. I.E. if a 1 millisecond burst of signal is transmitted on one half cycle (binary 1) then no signal should be transmitted on the next cycle, (binary 0). See Figure 3.

The Tables in Figure 4 show the binary codes to be transmitted for each House Code and Key Code. The Start Code is always 1110 which is a unique code and is the only code which does not follow the true complimentary relationship on alternate half cycles. .

HOUSE CODES					KEY CODES					
	H1	H2	H4	H8		D1	D2	D4	D8	D16
A	0	1	1	0	1	0	1	1	0	0
B	1	1	1	0	2	1	1	1	0	0
C	0	0	1	0	3	0	0	1	0	0
D	1	0	1	0	4	1	0	1	0	0
E	0	0	0	1	5	0	0	0	1	0
F	1	0	0	1	6	1	0	0	1	0
G	0	1	0	1	7	0	1	0	1	0
H	1	1	0	1	8	1	1	0	1	0
I	0	1	1	1	9	0	1	1	1	0
J	1	1	1	1	10	1	1	1	1	0
K	0	0	1	1	11	0	0	1	1	0
L	1	0	1	1	12	1	0	1	1	0
M	0	0	0	0	13	0	0	0	0	0
N	1	0	0	0	14	1	0	0	0	0
O	0	1	0	0	15	0	1	0	0	0
P	1	1	0	0	16	1	1	0	0	0
					All Units Off	0	0	0	0	1
					All Lights On	0	0	0	1	1
					On	0	0	1	0	1
					Off	0	0	1	1	1
					Dim	0	1	0	0	1
					Bright	0	1	0	1	1
					All Lights Off	0	1	1	0	1
					Extended Code	0	1	1	1	1
					Hail Request	1	0	0	0	1 [1]
					Hail Acknowledge	1	0	0	1	1
					Pre-Set Dim	1	0	1	X	1 [2]
					Extended Data (analog)	1	1	0	0	1 [3]
					Status=on	1	1	0	1	1
					Status=off	1	1	1	0	1
					Status Request	1	1	1	1	1

FIGURE 4

[1] Hail Request is transmitted to see if there are any X10 transmitters within listening range. This allows the O.E.M. to assign a different Housecode if a "Hail Acknowledge" is received.

[2] In a Pre-Set Dim instruction, the D8 bit represents the Most Significant Bit of the level and H1, H2, H4 and H8 bits represent the Least Significant Bits.

[3] The Extended Data code is followed by 8 bit bytes which can represent Analog Data (after A to D conversion). There should be no gaps between the Extended Data code and the actual data, and no gaps between data bytes. The first 8 bit byte can be used to say how many bytes of data will follow. If gaps are left between data bytes, these codes could be received by X10 modules causing erroneous operation.

Extended Code is similar to Extended Data: 8 Bit bytes which follow Extended Code (**with no gaps**) can represent additional codes. This allows the designer to expand beyond the 256 codes presently available. .

NOTE 1 . X10 Receiver Modules require a "silence" of at least 3 power cycles between each pair of 11 bit code transmissions (no gaps between each pair). The one exception to this rule is **bright and dim** codes. These are transmitted **continuously** with no gaps between each 11 bit dim code or 11 bit bright code. A 3 cycle gap is necessary between different codes, i.e. between bright and dim, or 1 and dim, or on and bright, etc.

NOTE 2. The TW523 Two-Way Power Line Interface cannot receive Extended Code or Extended Data because these codes have no gaps between them. The TW523 can only receive standard "pairs" of 11 bit X10 codes with 3 power line cycle gaps between each pair.

NOTE 3. The TW523 can receive dim and bright codes but the output will represent the first dim or bright code received, followed by every third code received. i.e. the output from the TW523 will not be a continuous stream of dim and bright codes like the codes which are transmitted.

A Square wave representing zero crossing detect is provided by the PL513/TW523 and is within 100 µs of the zero crossing point of the AC power line. The output signal envelope from the O.E.M. should be within 50 µs of this zero crossing detect. The signal envelope should be 1 ms (-50µs +100µs). See Figure 5. .

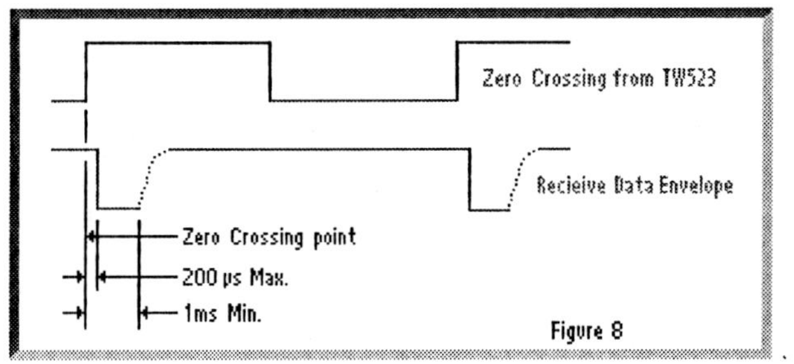

Figure 8

Opto-Coupled 60 Hz reference output (*from* the PL513/TW523) Transmissions are to be synchronized to the zero crossing point of the AC power line and should be as close to true zero crossing as possible. The PL513 and TW523 are designed to be interfaced to other microprocessor circuitry which outputs X10 codes synchronized to the zero crossing point of the AC power line. It is therefore necessary to provide a zero crossing reference for the O.E.M. microprocessor.

It is likely that this microprocessor will have its own "isolated" power supply. It is necessary to maintain this isolation, therefore the trigger circuit normally used in X10 POWERHOUSE controllers is **not** desirable as this would reference the O.E.M. power supply to the AC power line. It is also **not** desirable to take the trigger from the secondary side of the power supply transformer as some phase shift is likely to occur. It is therefore necessary to provide an opto-coupled 60 Hz reference.

An opto-coupled 60 Hz square wave is provided at the output of the PL513 and TW523. X10 codes generated by the O.E.M. product are to be synchronized to this zero crossing reference. The X10 code envelope generated by the O.E.M. is applied to the PL513 or TW523 which modulates the envelope with 120 kHz and capacitively couples it to the AC power line.

Opto-Coupled Signal Input (*to* the PL513/TW523)

The input signal required from the O.E.M. product is the signal "envelope" of the X10 code format, i.e.

High for 1 ms. coincident with zero crossing represents a binary "1" and gates the 120 kHz oscillator through to the output drive circuit thus transmitting 120 kHz onto the AC power line for 1 ms.

Low for 1 ms. coincident with the zero crossing point represents a binary "0" and turns the 120 kHz oscillator/output circuit off for the duration of the 1 ms. input.

X-10 code recieived from the AC power line.
START CODE
HOUSE CODE
FUNCTION CODE
START CODE
HOUSE CODE
FUNCTION CODE
START CODE
HOUSE CODE
FUNCTION CODE
"X-10 recieived" output from TW523

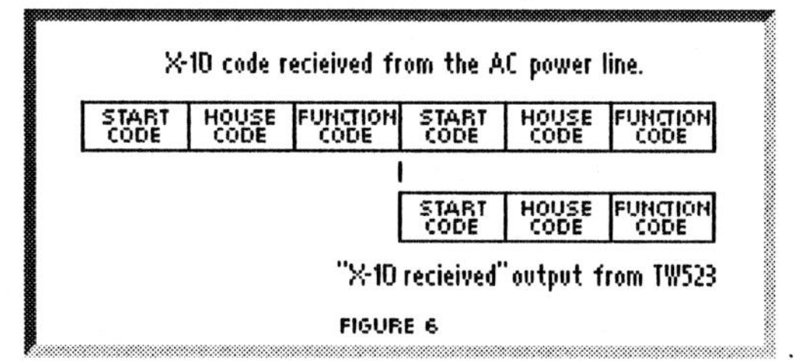

FIGURE 6

Opto-Coupled Signal Output (*from* the TW523)

The "X10 received" output from the TW523 coincides with the second half of each X10 transmission. This output is the envelope of the bursts of 120 kHz received. Only the envelope corresponding to the first burst of each group of 3 bursts is available at the output of the TW523. See Figures 6, 7 and 8. .

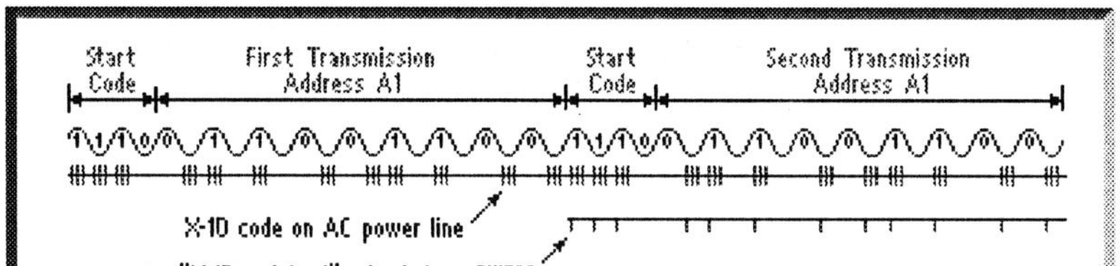

Start Code
First Transmission Address A1
Start Code
Second Transmission Address A1
X-10 code on AC power line
"X-10 recieived" output from TW523

Figure 7

Source: X10 WorldWideWeb pages

Lighting:

This part is actually a sample freebie I thought I'd throw in from another work I hope to release soon. In the near future I expect to produce a "scripting wizard" to give people an idea of where to go when they begin programming their Stargate. Consider this an example of what it contains!

While designing a lighting script is extremely easy, you have almost limitless options of how you might implement it. For instance, lighting routines can be interfaced with other scripted elements like doorbells, telephones, and (most obviously AND in this article) security systems.

Modes:

The first place to begin in your lighting design is to create several "modes" for your home. Each mode will determine distinct behaviors for the automation system, & the simple fact that they exist will make programming your lighting much easier, as you'll soon see. You can create as many or as few modes as you wish but you probably won't need more than about three major ones. Here, we'll call them "**HOME**", "**AWAY**", and "**VACATION**." I suspect you know what they mean, but just to make sure we're speaking the same language:

HOME is for how you want things to run when you're present.
AWAY is for when the home is unoccupied for short periods of time (i.e. when you're at work).
VACATION is for when nobody is around for days/weeks or more.

**Note: You'll find yourself creating other flags (modes) to run under HOME, AWAY, and VACATION. For instance, you may have a PARTY mode for when you're entertaining which runs concurrently with HOME mode.*

The "devices" that we'll create in the WinEVM software are simple "**Flags**." To begin, open your schedule (or click "**File|New Schedule**" if you don't have one yet). Click "**Define|Flags**" and name the three flags "home, away," and "vacation." The "description" is optional, but you should make a decision concerning the "initial state" of the flags. "Initial State" simply refers to how you want Stargate to set the flag immediately upon downloading the schedule. Since you'll want one of these flags (and only one) to be set at all times, and also since you'll probably be HOME when you do your download, I'd recommend that you set the **initial state** of the "home" flag to SET, and the others to CLEAR.

Click "OK" and now you're ready to add these to your schedule. Simple enough. Click "**new event**," name it HOME MODE and click OK (just make it a standard "if/then" event. By the way, you should read Stargate's manual to make sure you understand the different types of events). The "if" line in your schedule should now be highlighted. Click "**add|flag**", select "home" and check the "set" option. Click "OK."

Now highlight "then" in the event schedule. Click "add|flag", select "away" and check the "clear" option. Click "OK" and then repeat with the "vacation" flag. Your event should look as it does below.

```
EVENT: Home mode
If
  (F:home) is SET
Then
  (F:away) CLEAR
  (F:vacation) CLEAR
End
```

We now simply have to repeat the same process two more times. You should create two more events, one which sets the "away" flag instead of "home" (as above), and one which sets the "vacation" flag. Now what we have (see below) is a situation in which the setting of any one flag will automatically clear all the others. Bear in mind that we haven't told the system to DO ANYTHING yet - we're just preparing it to think clearly.

```
EVENT:  Home mode
If
  (F:home) is SET
Then
  (F:away) CLEAR
  (F:vacation) CLEAR
(End)
EVENT: Away mode
If
  (F:away) is SET
Then
  (F:home) CLEAR
  (F:vacation) CLEAR
End
EVENT: Vacation mode
If
  (F:vacation)is SET
Then
  (F:home) CLEAR
  (F:away) CLEAR
End
```

There are various ways to set the modes in your home. You can do it manually from a touch-tone phone, keypad, or X10 controller...

Or you can have it happen automatically, such as when you arm/disarm your security system.

For now, let's add the following schedule to manually implement the modes via touch-tone phone.

```
Event: Set Home Mode
If
  Telephone Seq:'^44' Received within 3 seconds
Then
  (F:home) SET

End
```

```
EVENT: Set Away Mode
If
  TelePhone Seq:'^55' Received within 3 seconds
Then
  (F:away) SET
End
EVENT: Set Vacation Mode
If
  TelePhone Seq:'^66' Received within 3 seconds
Then
  (F:vacation) SET
End
```

Note: Stargate can set your phones to operate by default on "intercom mode." When phones are on intercom, you do not get a dial tone just by picking up the phone (you have to program Stargate to give dial tone by pressing a digit, i.e. "9").

If you'll be locally using your telephones a lot to control your home, you may want to do this. This way you won't find yourself accidentally dialing the operator or 900 numbers, etc.

Below is a script you can use to accomplish this if you wish. It assumes that you're operating by default on intercom mode:

```
EVENT: Intercom auto switch to CO
If
  CO: Ring 1
Then
  Connect PHONE port to CO port
End
EVENT: switch back to intercom
If
  CO: Is ON Hook
Then
  DELAY 0:00:02
  Connect PHONE port to ICM port
End
EVENT: call out
If
  CO: Is OFF Hook
Then
  (F:TT off) SET
End
EVENT: TT off
If
  (F:TT off) is SET
Then
  Disable TouchTone System
Else
  Enable TouchTone System
End
EVENT: TT flag clear
If
  (F:TT off) is SET
  and CO: Is ON Hook
Then
  (F:TT off) CLEAR
End
```

By the way, notice the lines highlighted in red. If you don't have a device in your database, it will show up like this in your schedule. In this script, you'll have to **create a flag called "TT off." Notice that when the flag is set the TouchTone System is disabled. The reason for this is so that you don't send X10 codes with your phone when you don't mean to do so.**

Lighting Scenes:

What we want to do here is create a number of basic lighting scenes. These may be called up manually for a desired atmosphere, or as part of an automatic routine. We'll get around to implementing them later, but first let's create the scenes. For *your* purposes you would of course substitute your devices (lights) for the ones I show below.

Once again, we'll use flags (for different lighting scenes) and design them so that only one runs at a time. When one is set, the others are cleared. Before you can create the schedule below, you'll have to define the devices (flags, X10 lights) in the device database.

You might notice that once a scene flag is "set", it then is programmed to go into the **"idle"** state. This is a neat feature of Stargate that comes in handy from time to time. Here's how it's useful in this case:

Let's say you set "scene#1." As time goes by, people turn some of the lights on or off manually for whatever reason. Later, you wish to restore things to the "scene#1" status. *If the flag is still set, Stargate will not respond because it reasons that the flag is already set!*

This is where the "idle" state comes in useful. It simply changes the state of the flag so that if you wish to set it again, Stargate will see that you're introducing a change and take action.

```
EVENT: lightscene1
If
  (F:scene#1) is SET
Then
  (F:scene#1) IDLE
  (F:scene#2) CLEAR
  (F:scene#3) CLEAR
  (F:scene #4) CLEAR
  (F:scene#5) CLEAR
  X10: C-5 PRE-Set Level 100%
  X10: B-6 theater room PRE-Set Level 100%
  X10: B-16 PRE-Set Level 84%
  X10: C-1 PRE-Set Level 100%
  X10: B-7 famrm tbl lmps Set Level 100%
  X10: B-10 OFF
```

```
  X10: B-11 OFF
  X10: B-12 OFF
  X10: B-13 OFF
  X10: B-14 OFF
  X10: B-15 OFF
End
```

EVENT: lightscene2

```
If
  (F:scene#2) is SET
  Then
  (F:scene#1) CLEAR
  (F:scene#3) CLEAR
  (F:scene#4) CLEAR
  (F:scene#5) CLEAR
  (F:scene#2) IDLE
  X10: C-5 PRE-Set Level 23%
  X10: B-6 theater room PRE-Set Level 23%
  X10: B-16 PRE-Set Level 19%
  X10: C-1 OFF
  X10: B-15 Set Level 80%
End
```

EVENT: lightscene3

```
If
  (F:scene#3) is SET
Then
  (F:scene#1) CLEAR
  (F:scene#2) CLEAR
  (F:scene#4) CLEAR
  (F:scene#5) CLEAR
  (F:scene#3) IDLE
  X10: C-5 PRE-Set Level 77%
  X10: B-6 theater room OFF
  X10: B-16 PRE-Set Level 26%
  X10: C-1 PRE-Set Level 32%
  X10: B-7 famrm tbl lmps Set Level 50%
  X10: B10 PRE-Set Level 90%
End
```

EVENT: lightscene4

```
If
  (F:scene#4) is SET
Then
  (F:scene#1) CLEAR
  (F:scene#2) CLEAR
  (F:scene#3) CLEAR
  (F:scene#5) CLEAR
  (F:scene#4) IDLE
  X10: B-2 garage flouresc Set Level 100%
  X10: B-3 front porch Set Level 100%
  X10: B-4 kitchen chandelr Set Level 100%
  X10: B-9 stereo lamp Set Level 80%
  X10: D-1 ON
End
```

EVENT: lightscene5

```
If
  (F:scene#5) is SET
Then
  (F:scene#1) CLEAR
  (F:scene#2) CLEAR
  (F:scene#3) CLEAR
  (F:scene#4) CLEAR
  (F:scene#5) IDLE
X10: B-10 PRE-Set Level 100%
X10: B-11 PRE-Set Level 29%
X10: B-12 Set Level 90%
X10: B-13 Set Level 80%
X10: B-14 PRE-Set Level 81%
X10: B-15 Set Level 90%
End
```

Security:

At this point you'll need to have interfaced your security system with the Stargate. To do this you'll need to consult your security panel's installation instructions.

In the Security section I discuss how I used the Caddx NX8e panel, which connects directly to Stargate via a com port. But if I were to use another panel the security sensors would all have to loop through Stargate's digital inputs. I'd also have to run something from one (or more) of the security panel's aux outputs to Stargate to feed info regarding alarm/arm/disarm status.

For example, in my previous residence I used the Caddx 8920 Ranger & programmed the Aux output to send 12V when the security panel was armed. Then I simply ran a pair of wires from the Aux output on the Caddx panel to one of the digital inputs of Stargate. When voltage would flow, Stargate would know! What I present below assumes the use of such a generic security system.

Let me explain what's going on below in the **first event**. When Stargate's digital input recognizes that the security system is armed, it begins a 3 minute countdown and plays a voice recording reminding me that my ferret (Teddy) needs to be put up (you'll need to delete this or substitute your own). Obviously this was for my own purposes - the little goober kept tripping my alarm when I wasn't home.

After the 3-minute countdown a nested if-then event sets the AWAY mode *AS LONG AS* two things are (still) true:

1. The security system is still armed
2. The security system's "motion bypass" is not in effect (which would indicate that someone is still home - this is a feature which Stargate would also need to know about).

The second nested if-then event occurs when someone later returns home. In this case a couple of verbal warnings play over the speakers to remind me to disarm the security system before the alarm goes off.

The **second event** simply activates the home mode when the security system is disarmed.

The **third event** isn't really important. If I arm the security system before I go to bed, it's kind of nice to be reminded of it in the morning.

```
EVENT: automatic "away"
If
  "armed"
  and (DI:sec.armed) is ON
Then
  Voice:teddy's loose [Spkr]
  DELAY 0:03:00 Re-Triggerable
    If
      (D:sec.armed) is ON
      and (DI:motionbypass) is OFF
      and (F:home) is SET
    Then
      (F:away) SET
    Nest End
    If
      (DI:garage door) Toggles
        -OR-
      (DI:front door) Toggles
    Then
      DELAY 0:00:04
        If
          (D(:sec.armed) is ON
        Then
          Voice:SECSYS ACTIVE [Spkr,ICM]
        Nest End
      DELAY 0:00:10
        If
          (DI:sec.armed) is ON
        Then
          Voice:10 SECONDS TO ALARM [Spkr,ICM]
        Nest End
      DELAY 0:00:05
  Nest End
End
EVENT: security disarmed
If
  (DI:sec.armed) Goes OFF
Then
  (F:home) SET
End
EVENT: morning security reminder
If-Always
  (DI:motionbypass) is ON
  and Time is After 6:00 AM SMTWTFS
  and Time is Before 9:00 AM SMTWTFS
  and (DI:family PIR) Toggles
Then Voice:SECURITY SYSTEM ACTIVE [Spkr]
  DELAY 0:40:00
End
```

Now, suppose you set "away" with the intention of being gone for just a few hours, but - oh, I don't know - ended up in the hospital (*hope* you're wearing clean underwear). It would be helpful if your home could automatically go into "vacation" mode.

Below is a short one that I've not had the opportunity to fully test, but it should work. With this in your schedule, when you set the "away" flag a timer begins an 18-hour countdown. Once the 18-hour timer has completed, if the home is still set for "away", then "vacation" mode will automatically be set.

I'm leaving it this way to keep it simple, but **you might want to add another event to your schedule**. Suppose you come home after 17 hours (set home mode), and then leave for a short trip to the store - do you really want your home to go into "away" mode in another hour when the timer expires? Adding another event whereby the setting of "home" mode stops the "set vacation" timer would eliminate that possibility.

```
EVENT: automatic "vacation"
If
  (F:away) is SET
Then
  (T:set vacation) LOAD with 18:00:00
   If
     (T:set vacation) is Expiring
     and (F:away) is SET
   Then
     (F:vacation) SET
   Nest End
End
```

To sum up what's happening here: It's a convenient way of automatically setting "home", "away", and "vacation" modes (presuming you typically will arm your security system when you're gone).

Now that this is in place, all kinds of other stuff can happen automatically!

Below is a short script to allow you to arm/disarm your security system through Stargate. This particular script uses the telephone, but you can modify it to use the LCD keypad, X10, or whatever *(I know we're talking about lighting - I'm just throwing this in for free)*.

To prepare for this you have to find the "keyswitch" on your security panel (again, consult your documentation). Usually it works via simple contact-closure. On my Caddx it's momentary contact-closure to either arm OR disarm. So, a pair of wires from the keyswitch contacts to one of the relays on Stargate gives complete control! You can call in to set or disarm your home from anywhere in the world!

```
EVENT: remote arm/disarm
If
  TelePhone Seq:'^3333' Received within 6 seconds
Then
  (RELAY:remote arm) ON
  DELAY 0:00:01
  (RELAY:remote arm) OFF
End
```

(By the way, this doesn't work on my house!)

One more script below for your security needs: If you don't have a monitored system (i.e. you're using just a local siren), you can purchase dialers to call you or a neighbor. **OR you can program your own...**

Let me explain a couple of things. You'll need to interface a pair of Stargate's digital inputs with a voltage source from your security panel that activates when your alarm goes off. I happened to use the siren outputs, but you might use another set of Aux outputs, etc.

Secondly, I created a flag *("AlarmAcknowledge")* that's set when the alarm is violated. The script behaves as follows: Stargate calls me to inform me of the alarm violation *(unfortunately Stargate cannot perceive when someone answers its calls, so a little creativity is required)*...

The message is announced several times & SG then hangs up. If I haven't acknowledged the message during the call by pressing **"*"** *(see the second event below)*, Stargate assumes I didn't get the message, and it calls me again.

```
EVENT: Alarm Notification
If
  (DI:sec.alarm!) Goes ON
Then
  (F:AlarmAcknowledge) SET
  TelePhone Out:'^555-1212,,,'
  Voice:ALARM VIOLATED At address [CO,ICM]
  DELAY 0:00:02
  Voice:ALARM VIOLATED AT address [CO]
  DELAY 0:00:02
  Voice:ALARM VIOLATED AT address [CO]
  Voice:CAMERA ACTIVATD [Spkr]
  Voice:POLICE ENROUTE [Spkr]
  DELAY 0:00:01
  Voice:ALARM VIOLATED AT address [CO]
  DELAY 0:00:02
  Voice:ALARM VIOLATED AT address [CO]
  Voice:CAMERA ACTIVATD [Spkr]
  Voice:POLICE ENROUTE {Spkr]
  DELAY 0:00:01
  Voice:ALARM VIOLATED AT address [C0]
  TelePhone Out:',,+'
    If
      (F:AlarmAcknowledge) is SET
    Then
      DELAY 0:00:02
      TelePhone Out:'^555-1212,,,' Voice:ALARM        VIOLATED At address [CO,ICM]
      DELAY 0:00:02
      Voice:ALARM VIOLATED AT address [CO]
      DELAY 0:00:02
      Voice:ALARM VIOLATED AT address [CO]
      Voice:CAMERA ACTIVATD [Spkr]
      Voice:POLICE ENROUTE [Spkr]
      DELAY 0:00:01
      Voice:ALARM VIOLATED AT address [CO]
      DELAY 0:00:02
      Voice:ALARM VIOLATED AT address [CO]
      Voice:CAMERA ACTIVATD [Spkr]
```

```
    Voice:POLICE ENROUTE {Spkr]
    DELAY 0:00:01
    Voice:ALARM VIOLATED AT address [C0]
    TelePhone Out:',,+'
    (F:AlarmAcknowledge) CLEAR
    DELAY 0:00:01
    TelePhone Out:',,+'
  Nest End
End
EVENT: Alarm Acknowledge
If
  (F:AlarmAcknowledge) is SET
  and TelePhone Seq:'*' Received within 91 seconds
Then
  (F:AlarmAcknowledge) CLEAR
  If
    (F:AlarmAcknowledge) is CLEAR
  Then
    TelePhone Out:'+'
  Nest End
End
```

Random Lighting:

Now that we've created our major "Modes," light scenes, and interfaced with the security system, we can do some more advanced lighting control.

These are scripts you can use to give the appearance of activity in the home during the AWAY or VACATION modes (both are required to work properly).

The first script you see below requires you to create three new flags:

- "random lighting"
- "dead of night"
- "its' light"

It's a little difficult to explain everything that's going on (you might have to study this some), but basically if AWAY or VACATION mode is active, a "then-macro" (random lights) kicks in after dark.

After 1:00 a.m., a more subdued lighting scheme takes place (a more natural appearance of someone being at home).

Finally, there are a couple of events which add some automatic outdoor lighting at night even if HOME mode is in effect. In the third event below you'll see referenced a "then-macro." This is something that Stargate allows you to do to save space in your schedule. The macro itself is outlined in the last section below.

EVENT: random lights
If
  (F:vacation) is SET
  and After Sunset SMTWTFS
    -OR-
  (F:vacation) is SET
  and Time is Before 1:00 AM SMTWTFS
Then (F:random lighting) SET
Else
  (F:random lighting) CLEAR
End

EVENT: random lights2
If
  (F:away) is SET
  and After Sunset SMTWTFS
    -OR
  (F:away) is SET
  and Time is Before 1:00 AM SMTWTFS
Then
  (F:random lighting) SET
Else
  (F:random lighting) CLEAR
End

EVENT: random lights=then macro
If
  (F:random lighting) is SET
Then
  (THEN MACRO:random lights)
End

EVENT: dead of night
If
  Time is After 1:00 AM SMTWTFS
  and Before Sunrise SMTWTFS
  and (F:vacation) is SET
    -OR-
  Time is After 1:00 AM SMTWTFS
  and (F:away) is SET
Then
  (F:dead of night) SET
Else
  (F:dead of night) CLEAR
End

EVENT: dead of night2
If
  (F:dead of night) is SET
Then
  (F:random lighting) CLEAR
  X10: B - All Lights Off
  X10: B-12 ON
  X10: B-2 garage PRE-Set Level 81%
  X10: B-4 kitchen chandelr PRE-Set Level 90%
  X10: B-9 stereo lamp ON
  X10: B-16 PRE-Set Level 19%
  X10: C-5 OFF
  X10: C-1 OFF
End

EVENT: it's light
If
  After Sunrise SMTWTFS
  and Before Sunset SMTWTFS
Then
  (F:it's light) SET
Else
  (F:it's light) CLEAR
End

EVENT: random lights off
If

```
  (F:it's light) is SET
  and (F:vacation) is SET
    -OR-
  (F:it's light) is SET
  and (F:away) is SET
Then
  X10: B-All Lights OFF
  X10: B-All Units OFF
  X10: D-1 OFF
  X10: C-1 OFF
  X10: C-5 OFF
End
```

EVENT: HOME outside lights off in day

```
If
  (F:it's light) is SET
  and (F:home) is SET
Then
  X10: B-2 garage OFF
  X10: B-3 front porch OFF
  X10: B-4 kitchen chandelr OFF
  X10: D-1 OFF
  X10: B-9 stereo lamp OFF
  X10: B-16 OFF
End
```

EVENT: HOME outside lights on at dark

```
If
  (F:it's light) is Not SET
  and (F:home) is SET
Then
  X10: B-9 stereo lamp PRE-Set level 81%
  X10: B-3 front porch PRE-Set level 90%
  X10: B-2 garage PRE-Set Level 81%
  X10: B16 PRE-Set Level 19%
End
```

What we have next is the actual macro that's referred to in the third event above (random lights). It's really the heart of what takes place in the script above, executing a continual loop of lighting scenes.

The last two nested if/then events ensure that the script keeps repeating itself until the "dead of night" flag is set (above), or until "home mode" is set. If you wanted, you could then define another macro to run when the dead of night flag is set - maybe something more subdued.

MACRO BEGIN

```
DELAY 0:00:03
  If
    (F:home) is Not SET
    and (F:random lighting) is Not CLEAR
  Then
    X10: B-1 kitchen OFF
    X10: B-4 kitchen chandelr PRE-Set Level 74%
    X10: B-3 front porch PRE-Set Level 90%
  Nest End
DELAY 0:19:00
  If
    (F:home) is Not SET
    and (F:random lighting) is Not CLEAR
  Then
    X10: B-4 kitchen chandelr OFF
    X10: B-1 kitchen PRE-Set Level 77%
    X10: B-3 front porch PRE-Set Level 52%
    X10: B-8 study lamp Set Level 50%
  Nest End
```

```
DELAY 0:11:00
  If
    (F:home) is Not SET
    and (F:random lighting) is Not CLEAR
  Then
    X10: B-8 study lamp OFF
    X10: B-2 garage ON
    DELAY 0:00:17
    X10: B-2 garage OFF
  Nest End
DELAY 0:33:00
If
    (F:home) is Not Set
    and (F:random lighting) is Not CLEAR
  Then
    X10: B-4 kitchen chandelr PRE-Set Level 48%
    X10: B-8 study lamp Set Level 90%
    X10: B-3 front porch PRE-Set Level 90%
  Nest End
DELAY 0:19:00
  If
    (F:home) is Not SET
    and (F:random lighting) is Not CLEAR
  Then
    X10: B-8 study lamp Set Level 20%
    X10: B-3 front porch PRE-Set Level 48%
    X10: B-1 kitchen PRE-Set Level 100%
    X10: B-4 kitchen chandelr PRE-Set Level 23%
  Nest End
DELAY 0:08:14
  If
    (F:home) is Not SET
    and (F:random lighting) is Not CLEAR
  Then
    X10: B-2 garage ON
    DELAY 0:00:06
    X10: B-2 garage OFF
  Nest End
  If
    (F:random lighting) is SET
  Then
    (F:random lighting) IDLE
  Nest End
  If
    (F:random lighting) is IDLE
  Then
    DELAY 0:00:03
    (F:random lighting) SET
  Nest End
MACRO END
```

Infrared System:

I mentioned earlier that I pulled cat5 to each volume control location for the house audio, which puts my basic IR wiring structure in place.

In truth, you can put IR sensors any place that's convenient. For instance, there are "bullet" IR sensors that can be mounted in a ceiling or speaker, or surface-mount sensors that can be discreetly attached to the front of a television. Do it any way you wish, but keep in mind that you'll probably keep your costs down if you stick to the standard in-wall type.

Keep in mind, too, just where you'll need to access your automation system, and how (IR, X10, keypad, etc.). You'll want to have an interface in your TV or Theater room, and probably several other places as well, like kitchen, bedrooms, etc.

The three-wire connection is really very simple on a Niles IRR4D eye. At my IRP6 unit (located in the equipment room) I have a "12V, ground, and data" connection which corresponds to "1,2,and 3" on the sensors (in that order). I don't know why they don't use the same markings on both, but if you're using Niles ... well, now you know what's what.

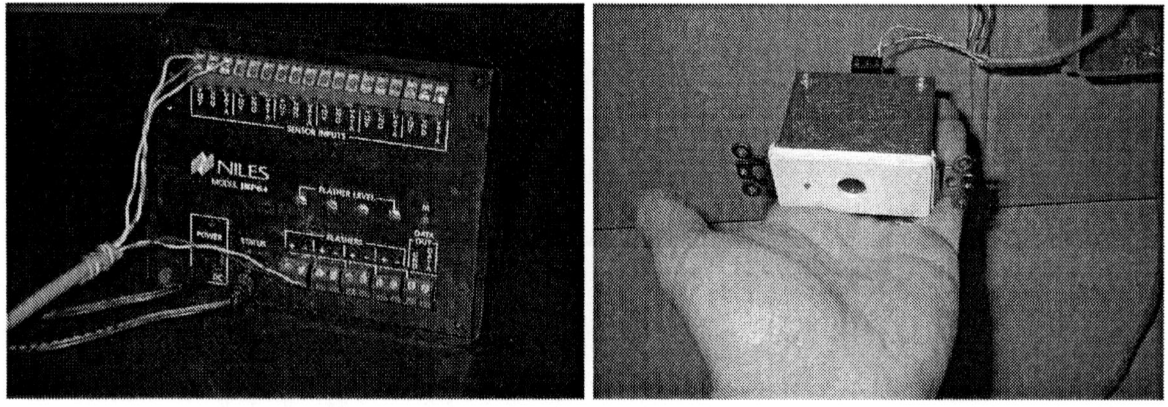

Niles IRP6 *Niles IRR4D*

In case I need to repeat what I've said before in Volume 1, when you aim a remote at the sensor (upper right picture) the info is relayed to the "junction box" (in this case the IRP6, upper left picture). The signal is then relayed to the flasher outputs (a two-wire connection) which are in turn hardwired to the emitters in front of the components to be controlled.

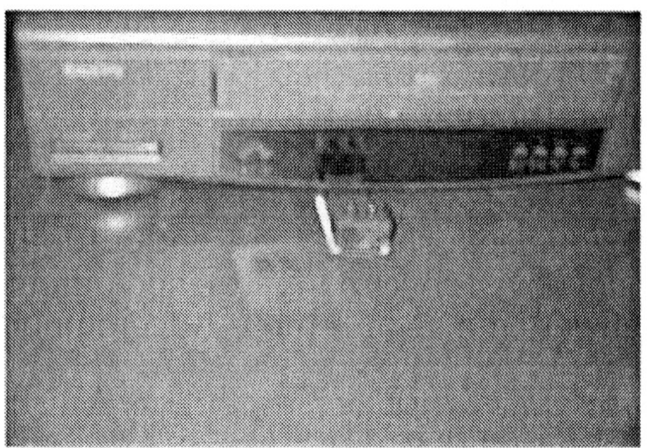

flooding flasher reaches several components

If you observe the pic of the IRP6 you'll see that there's an input for a status sensor. This was part of an original Niles design for their Intellipad keypad system. With a 12v power supply connected to a switched outlet on your stereo receiver, this provides the power status to the keypad.

Even though I'm using the Intellipad, it's only because I already happen to have it. If you're using Stargate, invest your money in the LCD96M keypad. The cost is about the same and functionality is far superior *(I'll talk more later about sensing the status of your equipment.)*

Niles Intellipad

So far we've looked at a simple standalone IR system. There are no intelligent capabilities built in. It's just a plain old stupid hard-wired infrared relay system.

But I actually have two IR systems. Stargate has its own IR interface (the IRXpander), but there's very little to do in the way of wiring. The two systems simply interface in the equipment room.

The IRXP provides Stargate with the ability to both send AND receive IR signals. If you look at the two pics below you can see how it's integrated with the Niles IR system.

Below on the left you see a shot of the front of the IRXP. There's a little IR sensor window over which I've attached an emitter from the Niles box. This effectively lets me send remote IR commands to Stargate.

The **pic on the right** shows a Niles infrared eye which I mounted in the equipment room just below the IRXP. It, in turn, has an emitter attached coming from the IRXP. Stargate can now send commands through the existing Niles network to all IR components.

Later on I'll get into more of this when it comes time to program. We're still just putting stuff into place.

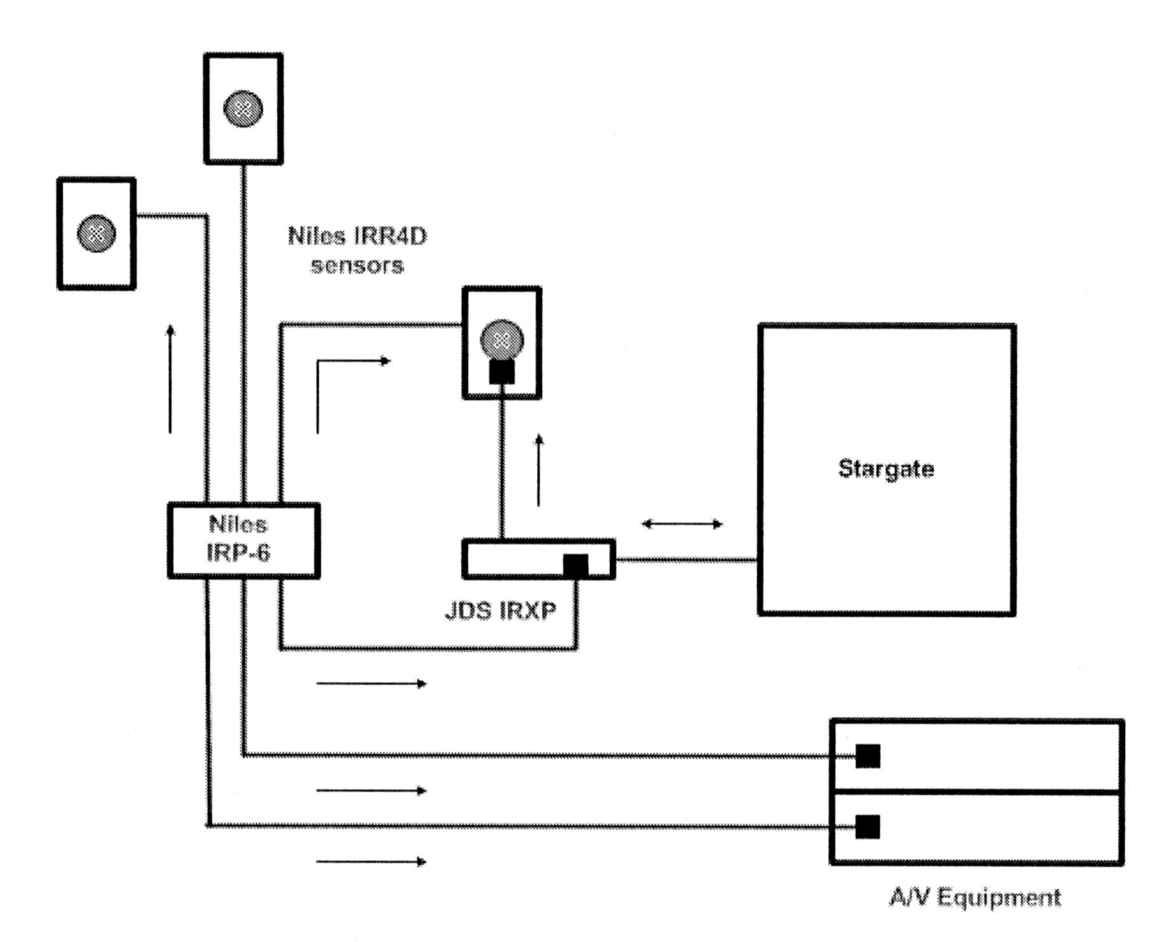

House Audio:

Now we get to something that's likely to be one of your favorite house features... at least if you use decent quality stuff.

As a byline, let me suggest that if you're limited on how much you can spend in this area, put your money into the speakers. You'll get the best return for your investment this way. Decent speakers can make a relatively cheap stereo sound pretty good. Conversely, cheap speakers will make expensive equipment sound like junk.

The notes that I took on my prewire were obviously important in locating my wiring. From the above notes in the Great Room, I knew the approximate measurement to the center of the chase (between the studs). I also thought it prudent to make sure by using a stud finder.

Often you'll find a balance needs to be struck between aesthetics and acoustics. You'll

find more discussion on this in Book 1, but here's how I handled things in my Great Room:

The best height for inwall speakers is at ear-level, but that might not be the best visually. At least that's what my wife said. So I dropped them to the level of the bottom of the picture window, lining the bottom edge of the speaker frame with the bottom of the wood trim.

Oh, and when you do this, don't trust your eyes to level the speakers. Make sure you use a "torpedo level." You can't impress your friends with your smart home if it looks gimpy!

In order to make things look as good as possible, I also adjusted the speaker placement within the stud chases so that they would each be exactly the same distance from the window. Maybe I'm being Mr. Obvious here - but it *does* require some careful measurements for both speakers (and knowing *exactly* where the studs are) before making any cuts with the drywall knife.

The inwalls that I used were RBH MC8's. Very nice in-walls. Below is a pic of the speaker before it was installed, showing the drywall cutout template and black plastic bracket which clamps to the back of the drywall. Some manufacturers will use wings which flip around to screw down tight against the back of the drywall.

If you want you can usually get mounting brackets that go in during prewire. This isn't my preferred method, though, because you'll have gaping holes in your wall until you install the speakers. However, some upscale speakers do in fact *require* "backboxes" to get the best sound, so your situation might vary from mine.

At long last, here's what the speaker looks like installed (minus the speaker grill):

At this point, I want to back up a little and talk for a brief moment about **the wiring** again. I wanted to have subwoofers installed in several rooms. This had to be taken into consideration during prewire, because the sub needs to be installed in-line between the volume control and speakers.

Once I chose the location for my sub, I ran 4-conductor speaker wire *from the volume control* to the sub's inputs, and the two 2-conductors then proceeded from the sub outputs to the speakers. This accomplishes the following:

The sub's cross-over filters off the lower frequencies and passes them on to the subwoofer. It sends everything else on to the speakers (sparing them the burden of trying to reproduce the heavy stuff). The sub's volume is also controlled by the volume control this way. Be aware that there are a few subwoofers on the market that only have line-level inputs & do not have speaker-level inputs. They wouldn't work in this scenario.

Here's a view of the volume control's connections. Pretty straight-forward, really. Just be sure that you observe + and - connections all the way from the amplifier to speakers.

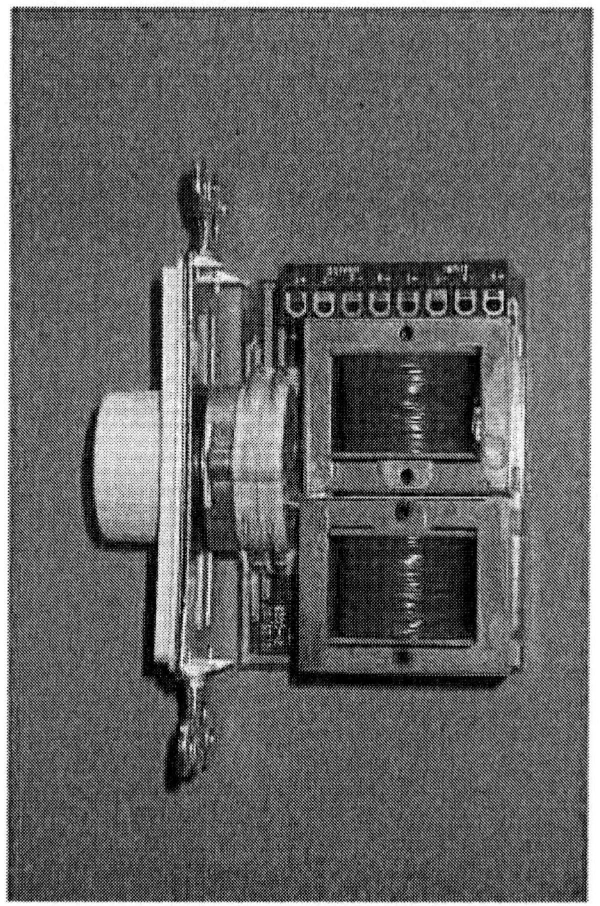

volume control - 4 connections at top for "amp in", and 4 for "speakers out"

Please don't buy the cheapest thing you can find. Decent quality volume controls will last longer, they won't make popping noises as they wear out, and will provide smoother increments in volume. Expect to pay $40 to $50 apiece.

You'll pay a little more if you buy "impedance-matching" volume controls, but they will obviate the need for a speaker-selector box. Personally, I prefer to use the box and NOT the impedance-matching VC's. That way I don't have to readjust every one of my volume controls as I add rooms to my system (you have to set the number of speaker pairs on the impedance-matching VC's).

I can also turn rooms on or off from my speaker-selector box. What you see below is a Russound 4-pair box. It has a speaker-level input for the 4-conductor coming from the amp, and outputs for 4 rooms of audio.

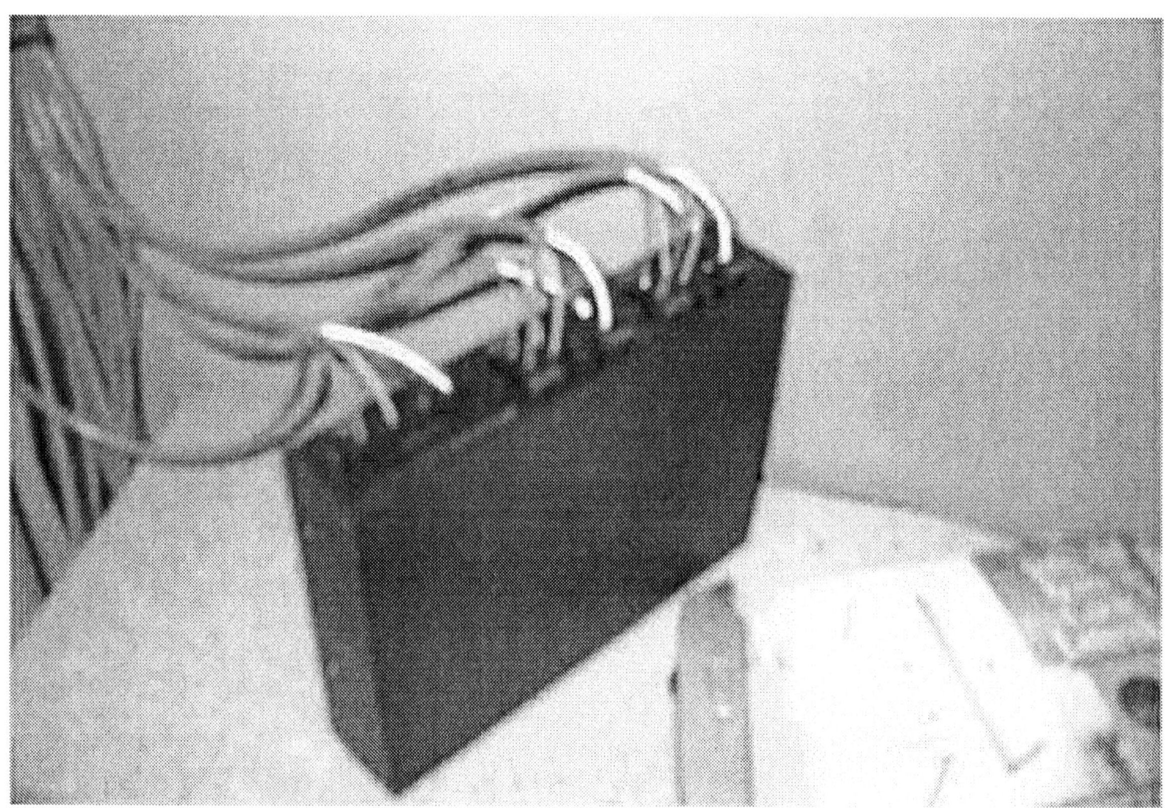

Every stinkin' time I took this picture it was blurred!

Interfacing house audio with Stargate:

Except for the IR interface between the IRXpander and the Niles IR system, the only way Stargate touches my audio system is through the speaker wire. Since Stargate has a speaker level output it made sense to me to make the house audio speakers accessible to it.

While there are several ways to do this, I chose what I thought would be the simplest method. Coming OUT of my audio amplifier, the 4-conductor passes through four sets of relays on Stargate's board. There is also a 4-conductor coming from SG's speaker output going into the relays, and a 4-conductor returning from the relays to the speaker-selector box.

In detail, each of the four relays consists of 3 terminals: NC, Common, and NO. The wiring is done as follows:

1. ***NC (normally closed):*** *wiring from audio amp goes here*
2. ***Common:*** *wiring TO speaker selector box goes here*
3. ***NO (normally open):*** *wiring from SG's on-board amp goes here*

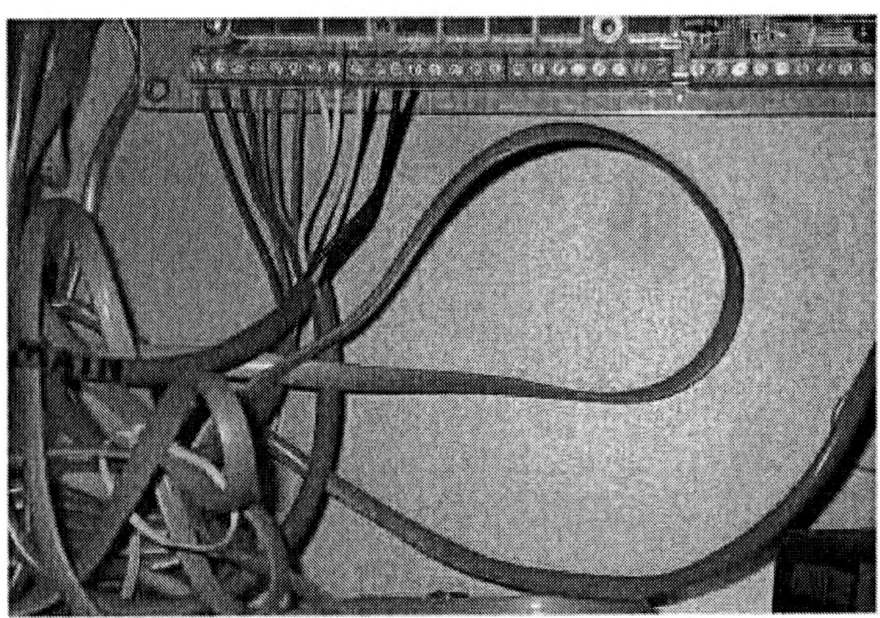

relays allow me to switch from one audio source to another.

The one issue that had to be dealt with was the fact that Stargate has a monaural output. Actually the main problem was that there was a *single* speaker output that needed to be played over a house stereo (two speaker) system.

A little trick which I believe is documented in JDS's Application notes is to wire in series the two speakers which are to be connected to Stargate's speaker output. It can be done as depicted below, being careful to observe polarity at the splice (one speaker's "+" to the other speaker's "-").

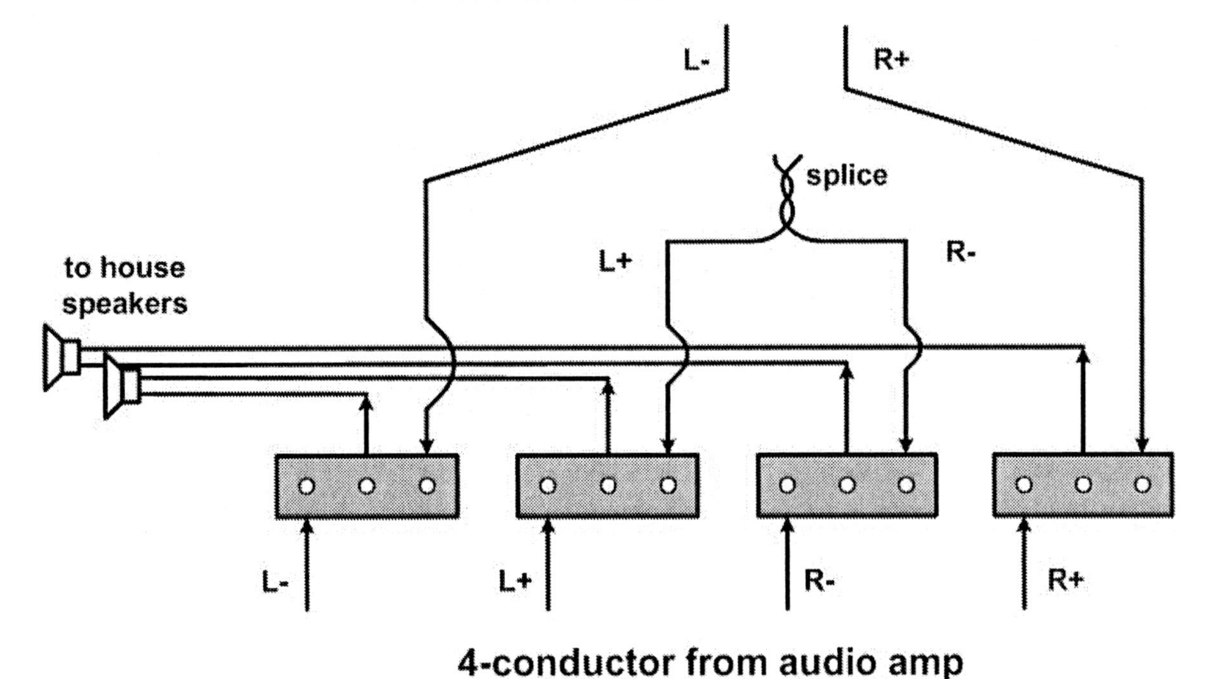

With this arrangement, music can be interrupted with Stargate voice prompts/ sound effects when necessary. For instance, when the doorbell or telephone rings, etc., I can program the relays to switch over to a Stargate announcement.

> Something you might wish to concern yourself with is the fact that your stereo might be averse *to NOT* having a place to route its output (when the relays are switched over to SG).
>
> It might be advisable to program an IR "mute" command to execute during that period of time.

Below are some events that implement the Stargate/Audio system interface. I begin by designing a couple of macros that handle the relay switching between my two audio sources:

Macro: Stereo speaker switch

```
MACRO BEGIN
  (RELAY: R+SPKR) OFF
  (RELAY: R-SPKR) OFF
  (RELAY: L+SPKR) OFF
  (RELAY: L-SPKR) OFF
MACRO END
```

Macro: Stargate speaker switch

```
MACRO BEGIN
  (RELAY: R+SPKR) ON
  (RELAY: R-SPKR) ON
  (RELAY: L+SPKR) ON
  (RELAY: L-SPKR) ON
MACRO END
```

Next a few simple ways of using the macros in my schedule:

EVENT: Stereo status

```
If
  (DI: Stereo power) is ON
Then
  (THEN MACRO: Stereo speaker switch)
Else
  (THEN MACRO: Stargate speaker switch)
End
```

EVENT: if Home

```
If
  (F: HOME) is SET
Then
  If
    (DI: Stereo power) is ON
  Then
    (THEN MACRO: Stargate speaker switch)
    DELAY 0:00:03
    (THEN MACRO: Stereo speaker switch)
    Nest End
DELAY 0:00:01
Voice: HOME MODE (Spkr,CO) THEN MACRO: Home Mode
```

Security:

Every job has its headaches. Seems like I ran into my share when it came to the security system. Caddx isn't new to me, but most of my experience has been with their Ranger series. However, here I'm using the NX8E - tailor made for Stargate with a direct serial connection between the two.

Caddx makes DL900 software (don't think it's available to end-users) for Windows-based programming and downloading to the panel. For some unknown reason (and I mean *nobody* can figure it out), it was a mistake for me to use it.

The first time I attempted my download it disconnected about halfway through. I then found myself locked out of the panel. Couldn't get in through the software OR from the keypad.

Don't know why. Never changed the default access code, or the default download access code. It just changed all by itself.

After spending some time on the phone with Caddx, they concluded that I had a bad panel.

I sent it back.

They sent me another one.

It happened again and I felt really stupid.

I no longer use the DL900 software (at least not via modem).

Since the security system is such an integral part of the whole design, you can see how this slowed down the installation.... But I have digressed enough!

When you're ready to install your panel & connect all your wiring, you need to choose for yourself how you'll be allocating your zones. The NX8E has eight zones out of the box, but it's expandable (you have to purchase additional hardware "modules"). I found that eight zones was enough for me, though I did have to group a number of sensors together.

For instance, even though I *did* home-run all the window sensors, it made sense to run a number of them in series & place them all on the same zone (see below, right).

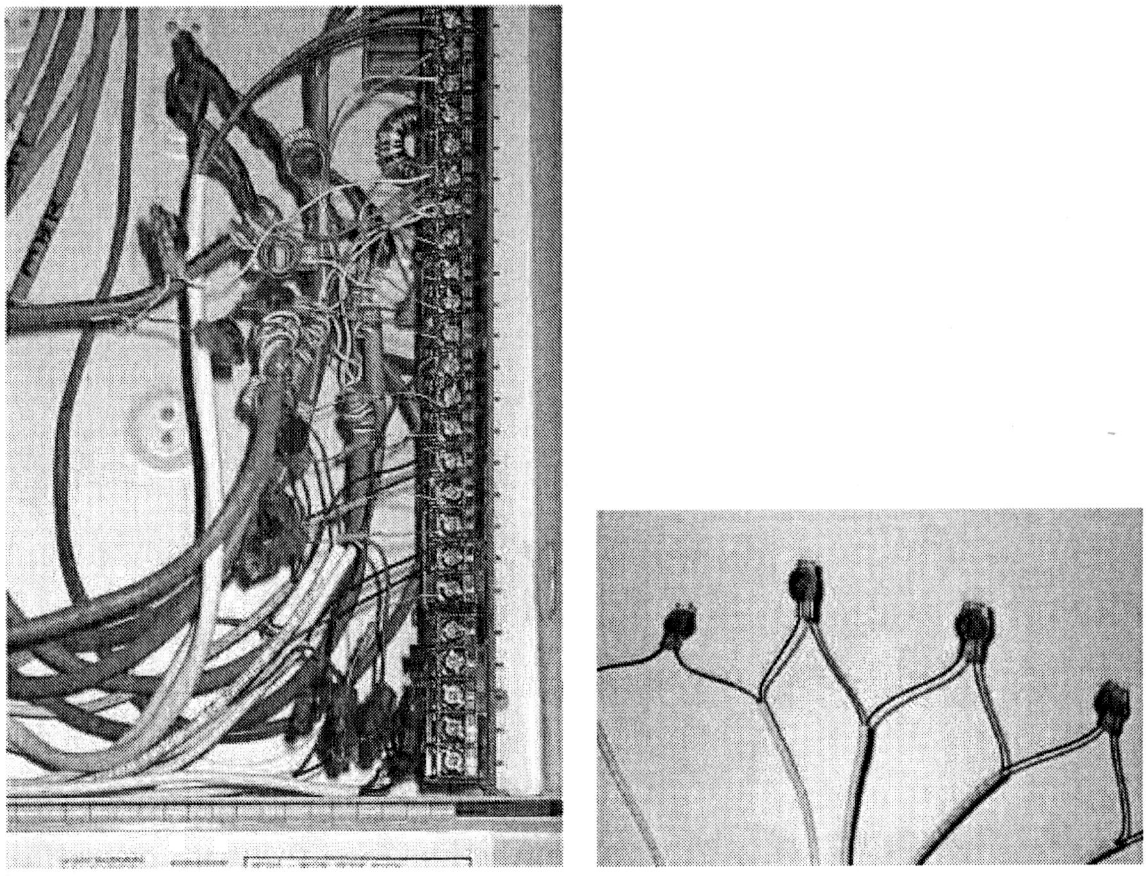

inside panel *windows wired in series*

As I said, this was the first time I programmed an NX8E, and it has the potential to be a little more complicated than the old Caddx Ranger series. In the end I found that, as a local system (no remote monitoring) there was really very little programming required. But I had to sift through a lot to figure that out.

One thing that threw me was a comment in the manual referring to the ability to obviate the need for EOL resistors. This didn't seem to be such a good idea (since I couldn't get the panel to recognize N.C. loops without them!). I might very well have been doing something wrong, but no matter. I'm probably better off with the EOL resistors.

Once I put the resistors on each zone (in series), programming became much simpler. With all my connections made, the NX8 was ready for arming with only its default settings when I plugged it in. However, I needed to do a little customizing... as follows:

Programming from the LCD Keypad:

The Keypad: Though you don't *have to* program your keypad, I chose to do so. Most of the default settings are fine for an eight-zone system, but I found it better to read "Front Door" on the keypad than "Zone 1" when there's a fault. However, if it makes no difference to you, you don't even have to bother with this.

I'm not going to hold your hand at this point since the manual is pretty self-explanatory when it comes to programming the keypad (mainly text on the display), but I will say this:

The manual instructs you to use the "master code" (default 1-2-3-4) at times, and at other times it tells you to enter the "program code" (default 9-7-1-3). You'll have to watch for this (and realize that they're not the same). It might be obvious to other people but I didn't notice at first & it really slowed me down.

The Panel: If you've never done a security system before, your first glance at the installation manual can be a little overwhelming. If you're actually going to be monitored by a central station you should probably get professional help here.

However, if it's a local alarm with a simple siren you can do this (*please understand that I assume no liability for your abilities!*). Get very familiar with the first few pages of the manual, especially pages 9-11 ("programming the NX8-E control"). This is where it tells you how to get around in programming mode.

To begin you enter ***8** at the keypad. Here the manual says a bunch of LED's should flash on the keypad. If you pay close attention you might notice that this is only if you have an LED keypad (and I hope you're using the LCD!). So you can ignore that comment in the manual.

Next you enter your program code (remember? default **9-7-1-3**), and the keypad should then ask you to specify the *device address* you wish to program. This is actually asking which "module" you'll be programming, which in this case is the panel itself (**0#**).

At this point you'll need to have determined your zone types (pages 18-19). Read over these to decide what should be what. As a for-instance, you'll likely set your entry doors as zone ***"type 3*: entry/exit delay type 1."** The built-in features of this zone will do the following:

When you set the alarm you have 60 seconds to vacate the premises before it arms. If you don't exit (i.e. open and close a zone type 3 door) within that period of time, the panel will assume you are staying home and automatically bypass your interior motion detectors (**zone *type 5***). Also, if you're returning home and you open the zone type 3 door, you'll be allowed 30 seconds to disarm the system.

The default 60 and 30 second timers are configurable in other programming locations, but seem to be appropriate for most people. Anyway, you'll find a large choice of zone types in the manual which should fit just about any need you might have.

Keep this in mind, though: Zone types 17, 18, and 20 are for wireless zones only!

Also, the siren output is *monitored*, meaning you'll need to make sure you have either a siren connected (I assume you would in the end anyway) or else an EOL resistor across the siren outputs. Otherwise, you'll get error messages on your keypad & the system *will not arm*.

After you've entered **0#** to get to the panel device location, you're ready to program your zone types into the panel. Go to location 25 ("**25#**") and you'll find eight "segments" corresponding to your eight zones. Enter your zone type for each zone, advancing through each one by pressing "*****."

When you're done, press "*" one last time and hit "exit" twice to get out of programming mode. You're essentially done, though some modifications may be necessary for your system (depending on your design).

Whether you change the default *Go To Program Code (9713)* is up to you. Just never, never, never, *not even* ***ever*** forget what you set it as. The only way back into programming mode would be via the download software.

You *will* probably want to change the default *Access Code (1234)* to something else. This is the code that arms/disarms the system.

Connecting to Stargate:

The NX8e has a built-in serial connector, but to interface with Stargate a small cable is required (*Caddx part #p0003*). It's only about $6 to $10, depending on your sources, but it will save you the additional configuration (and expense) of the 584 that you'll need if you have one of the other NX series panels.

The connection is very simple. Just make sure the red stripe is to the back of the panel when you install it. The other end of the cable connects to a DB9 to RJ11 adaptor & data cable, which in turn plugs into the COM2 or COM3 port on Stargate.

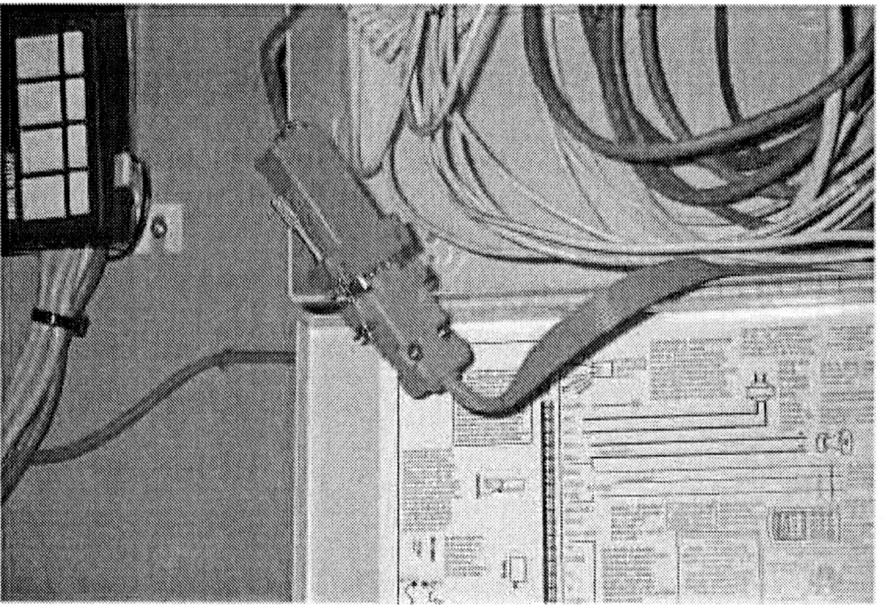

Caddx p0003 connector connects to DB9 adapter

> You might find that you're short on the required data cable & DB9 connector. Making the data cable is easy. You just need the 6-pin RJ12 connectors & crimpers. Then make the pinouts identical at both ends. Simple.

There are some setting that need to be made in the Caddx panel at this point *(by the way, it's recommended that you get the security system up and running correctly before you attempt this)*. If you need help getting into the panel, review the section on security or see the Caddx manual.

The settings are listed in a PDF which can be found at ***www.jdstechnologies.com/download/caddx.pdf.*** Other updated materials can also be found at their website under Appnotes.

One last thing to mention (which I hinted at earlier): I actually have a couple of zones connected to my Stargate instead of the Caddx panel. There were a couple of reasons for this. First, to simply save space & money (I didn't want to have to expand my number of zones on the Caddx. Second, these two zones aren't that important from a security standpoint.

I have an outdoor PIR at the front door which has the purpose of alerting me to movement outside. This device takes a 4-wire connection: two wires to power the PIR, and two wires to form the loop. I used the Caddx to power the PIR, but ran the loop through the digital inputs on Stargate.

The other zone in question monitors my garage door, a simple NC contact that also runs through SG's digital inputs.

This actually raises an issue I haven't discussed yet regarding the **digital inputs**. If you examine these, they have **jumpers** which can be set in one of two positions. In the "voltage" position, Stargate is looking for voltage from an exterior source (doorbell, etc.).

In the "switch" position, Stargate actually puts a small load of its own on the loop. I needed to use the "switch" position to put voltage on these two zones in question. It's ***important*** that you use these jumpers correctly!

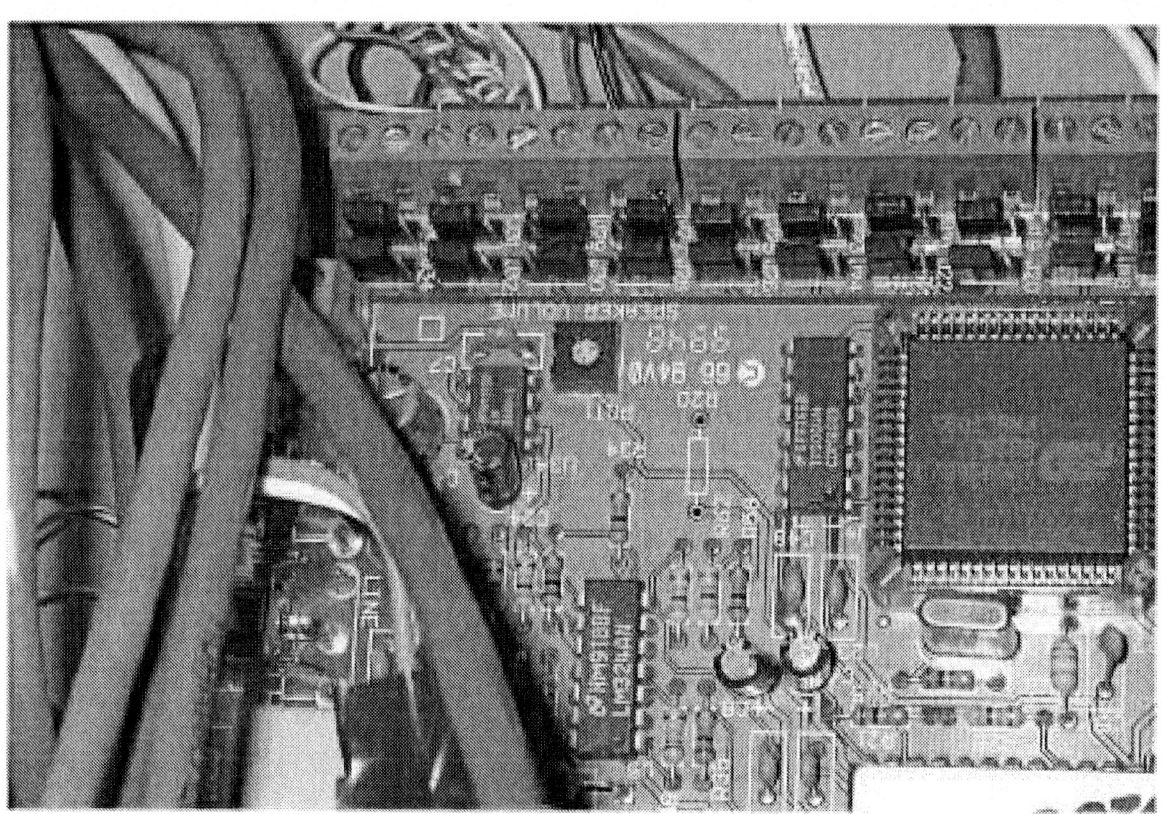

jumper pins on digital inputs (top edge of picture)

There really isn't much more for me to say about physical connections for security... because that's all there is to it!

Now it's just a matter of programming events. If you'd like you can jump back to the Lighting section to see some sample scheduling.

The JDS .pdf I mentioned earlier also has some examples concerning how to configure the LCD keypad for use in security.

Cameras & Webcams:

This not so bad. Easier than Caddx!

As I said before, cameras vary, but let me address the two main kinds that I used: outdoor weatherproof cameras, and indoor webcams.

Outdoor weatherproof camera: This sweet little thing is a Pro Video camera. If you don't want to spend a lot of money you can stay with black/white. Assuming that you'd like low-light resolution, b/w cameras work better than color anyway.

outdoor camera - I need to clean up the cables!

I have a camera currently mounted facing the front door. Night situations are dealt with by (HEY!) using lights so I can see who's there.

I've programmed my front porch lights to provide low-level illumination at dark. Besides that, a doorbell press or motion will trigger the lights to come to full brightness, allowing me to see who *or what* is out there. Is it just another pesky raccoon?? I could swear these things are bold enough to ring the doorbell.

This cam runs over coax. Power is delivered from a 12V transformer via a 2-conductor cable (though I ran an abundance of wiring: Cat5). You can see (below left) that I simply spliced my cabling into the power supply wiring with "Scotchlocks" *(Mr. Obvious says: the power supply wire wouldn't reach all the way to the camera, which is why I had to lengthen it).*

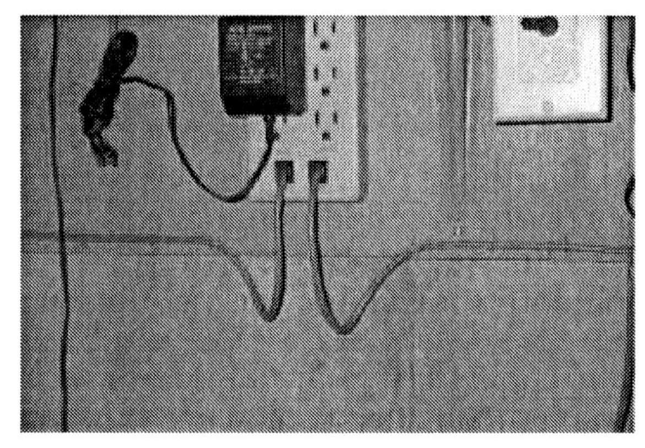

power supply splice

video modulator atop IRXP

The other end of the coax runs into my ChannelPlus modulator. A length of coax then comes out of the modulator and travels to my entertainment center, where it goes into one feed of a "combiner." The other feed into the combiner is from the output of my VCR. The output from the combiner then goes straight to the TV (see the diagram below).

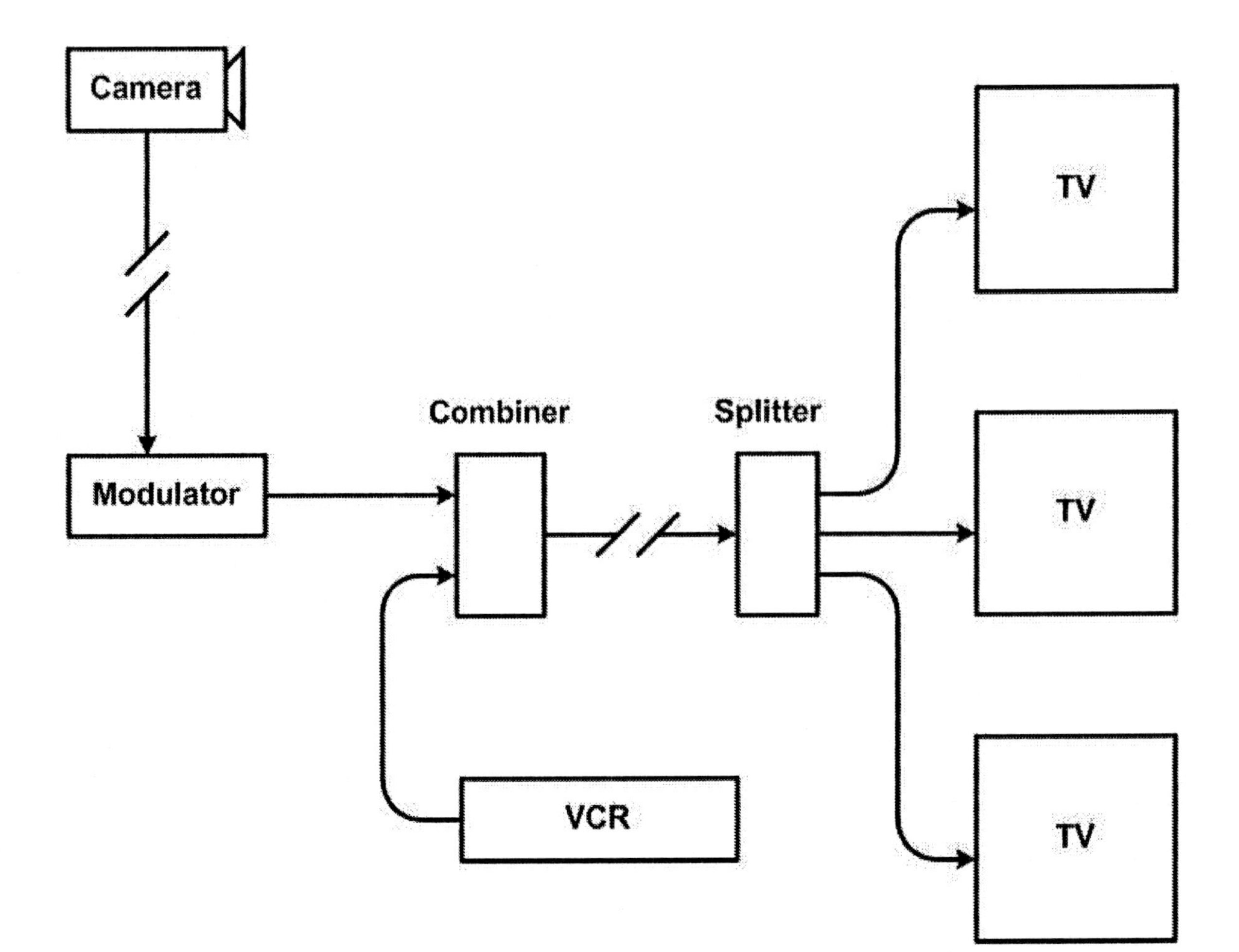

This is surely not the only way of doing this. You might not even need to use a combiner if your TV has an additional coaxial "antenna input." Basically it's just a question of where you want to inject the camera signal into the cable system.

For instance, I'm considering doing things a little differently since my TV's are always on channel 3 anyway (for satellite). I could just combine the camera signal BEFORE it enters the VCR and distribute the VCR signal to all TV's.

This way if I leave the VCR's tuner engaged at all times and program it to change to the camera channel upon doorbell presses, motion, etc. - it will automatically distribute the camera's image to all TV's with a single IR command.

Setting the modulator's channel to whatever you wish is usually pretty simple. You'll just need to follow the directions that come with your modulator. By the way, consider how many sources you'll want to distribute/modulate before you make your purchase. If you have multiple cameras consider a multi-channel modulator. One neat little device at a good price is the **NetMedia Triple Play** 3-channel modulator.

When all is done you should be able to turn to the appropriate channel on your TV (or VCR in my case) and see the camera's view. If you're having mysterious problems, check whether your TV is set to CABLE or ANTENNA. It could make a difference.

Webcams: I'm using two different kinds of webcams: Xanboo and Ivista. My Ivista package included 2 color cameras. One I've mounted in the basement right on the wall in the equipment room. The other is upstairs in my loft. My single Xanboo camera is in the Great Room.

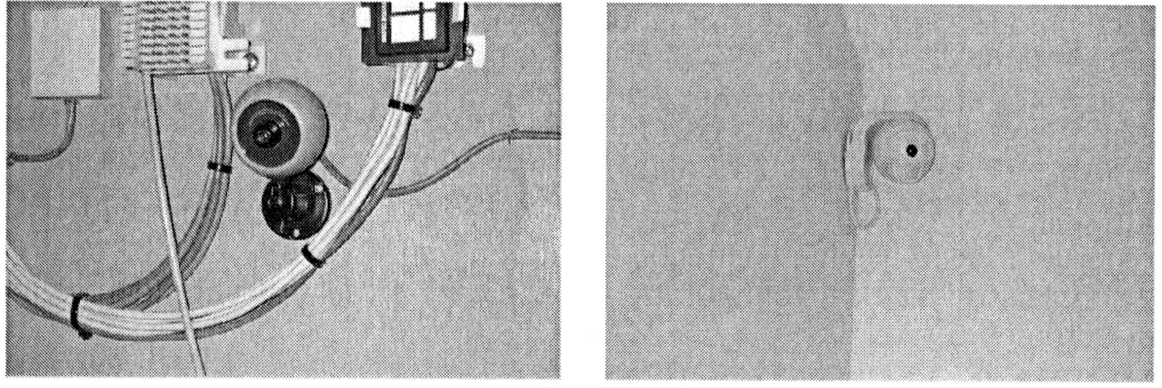

Ivista camera

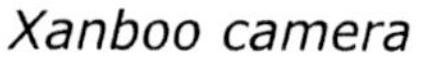

Xanboo camera

At the very beginning of the book I said you'd get to see me work through all the problems & bugs (less the hurled tools & profanity). Well, I created a little issue for myself (almost) when I prewired:

Originally my PC was going to be stationed upstairs in the loft. Since Xanboo has a 9-pin pre-made cable (100'), I ran it to the loft PC location. But as I've since said, my main PC is now in the equipment room where it really ought to be anyway.

Since I do in fact have a second PC in the loft, I've simply decided to use *it* to manage my webcams & let the "main" PC (equipment room) handle everything else. Problem is, I didn't run enough cabling to the loft to handle both Ivista webcams (really should follow my own advice and run much more than I think I'll need).

"Why didn't he?", you ask ("since he thought that the loft PC was going to do everything")?

I don't remember....

But I did run network wiring and a Cat5 for my PC-to-Stargate connection, which means:

Basically, I'm golden! One of my Ivista cams is going to be in the loft anyway (no additional wiring required there). I just needed another Cat5 back to the equipment room to the other Ivista cam, and since I won't be using the Stargate wiring which goes to the loft, I can use that for the webcam.

Lucky me. Otherwise I'd have to run new stuff (hate that).

Anyway, the nice thing about the Ivista software is that it can accommodate other webcam manufacturers. The Ivista "switcher" (kind of like a hub) connects to my loft PC's serial port, and the Xanboo unit utilizes a USB connection - but they can work together! The Ivista software will (should) accept other webcams. Nice.

I'm going to refer you back to the Data Network chapter for instructions regarding Ivista cabling connections. If you go this route you'll be using a simple straight-thru configuration (i.e. pin-outs to RJ45's are identical at both ends).

What I'll eventually be able to do with the Ivista software is monitor for movement (built-in motion detection) when the home is empty. I can FTP still shots (to one of my personal web sites) which are captured when movement is detected. I can also use the provided video-website for transmitting live video and audio.

If you don't understand how to "FTP", etc., don't worry. The software petty much walks you through it. Other things the software can do is send automatic emails, launch programs, etc. It's fairly versatile stuff.

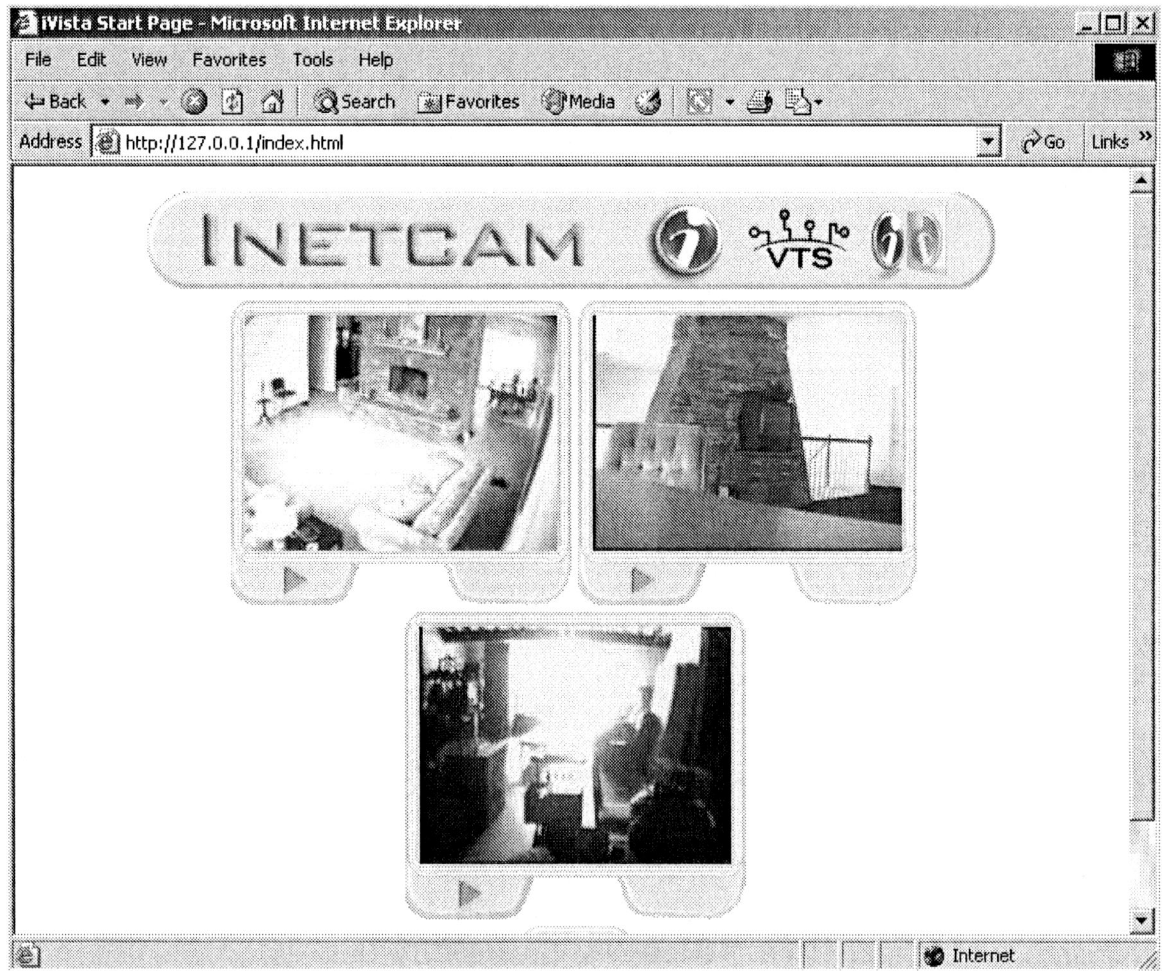

screenshot of my personal Ivista webpage

Integrating the two PC's & Webcam software:

At first I was baffled about how I was going to do this. As I said earlier, my Ivista webcams are managed by the PC in my upstairs loft, which in itself presents a problem in integration.

My design calls for the Ivista software to launch when my security system is armed ("away" flag is set on Stargate). That way, my home is covered by video surveillance and any movement will trigger an upload of video to one of my websites. But how to have Stargate launch the software on that remote PC? I couldn't just map a network drive & drop a shortcut into my schedule.

A query for help to the JDSusers forum (**www.jdsusers.com**) brought back a number of replies. The one that did the trick for me was a reference to **www.sysinternals.com**. There I found an abundance of freeware utilities, two of which met my exact needs:

The first one is called **Psexec**, a tiny little 120 K download which I placed ***into the root directory of my Stargate PC*** in the equipment room. After some fooling around, I found that I could launch the software on my loft PC from a command prompt on the SG PC. The command looked like this:

*psexec \\192.168.1.2 -u administrator -p ****** c:\progra~1\inetcam\programs\ivista.exe 1*

Let me interpret this for you: The command starts with "psexec" (the name of the utility) followed by the name or IP address of the remote computer (loft PC). Since the loft is a Win2K machine I have to provide a username (-u administrator) and password (-p *******[sorry I'm not telling]*). This is then simply followed by the path to the executable file I wish to launch.

Once this was possible, I created a batch file that I could include in my schedule. If you don't know how to create a **batch file**, it's a whole lot easier than you might imagine (as follows):

Creating a batch file: Open notepad and type in the same command that you'd enter at the DOS prompt, with one tiny exception. Instead of starting the command with "psexec", you'll need to provide the path to the psexec file. In my case, it reads "c:\psexec ..."

Under the File menu, select "save as." When the window pops up asking you where you'd like to save it, select your Stargate directory. Also, rename the file with a ".bat" extension, and select "all files" in the Files of Type dropdown menu. Save the file and you're done!

The WinEvm event looks something like this:

```
IF
  AWAY flag is set
THEN
  ASCII out:'&&ivista.bat'
```

This is how you launch a program from your Stargate schedule. The "&&" tells Stargate that what follows is an executable file.

NOW - I have to get into *this* scene for you: when psexec launches Ivista on my loft PC, it launches it *in the background*. I don't actually see the software running on the desktop at all, which means I can't just click the little "x" to close Ivista when I get home.

What I had to do was use the Windows Task Manager to identify the processes that run when Ivista starts up. I can manually shut them down from Task Manager, but that would be an obvious pain to do all the time. So I downloaded another neat little utility from www.sysinternals.com called **Pskill**.

Everything is essentially configured the same way, except that in the batch file I name the processes I want to kill rather than the path to an executable, as follows:

c:\pskill \\192.168.1.2 -u administrator -p 031976 IWS.exe
c:\pskill \\192.168.1.2 -u administrator -p 031976 ivista.exe
c:\pskill \\192.168.1.2 -u administrator -p 031976 inetcamserver.e
c:\pskill \\192.168.1.2 -u administrator -p 031976 ivistacapture.e
c:\pskill \\192.168.1.2 -u administrator -p 031976 inetmotdet.exe

If the "home" flag is set, Stargate runs "pskill.bat." Man, this is so cool....

HVAC:

Obviously I don't know what kind of heating/cooling system you have in your home, but mine isn't terribly complex. But whether you have multiple furnaces & a/c units or just one, the RCS units will connect in pretty much the same way.

I have two gas furnaces and two a/c units. Consequently, I have sort of a two-zone "out of the box" HVAC system (two thermostats). While it's possible to create more "zones" by installing electronic dampers, I found it unnecessary for my home. However, I'll go over this approach shortly to give you some ideas if it's something you'd like to do.

The only wiring that's really necessary for this is the RS485 (cat5) loop. I just had to make sure that it popped out somewhere near each of my furnaces (see below).

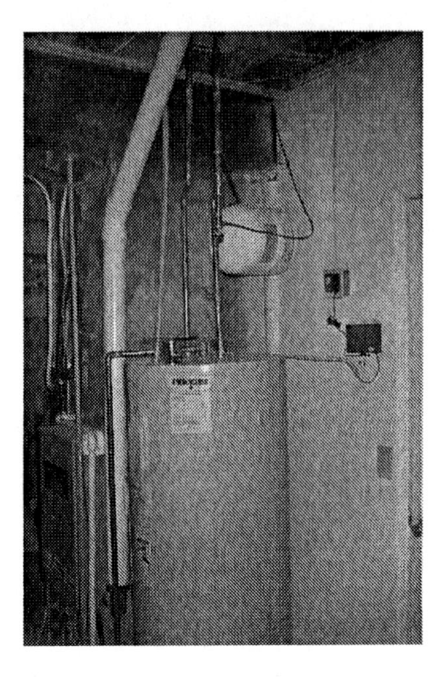

(Try to ignore the water heater. It has nothing to do with this other than to block your view of the furnace.) Anyway, on the right of the above pic you'll see the RCS control unit mounted on the wall next to the electric. Beneath it is the RS485 wiring and also a cat5 cable that interfaces with the thermostat wire.

It's a little easier to see in this next pic. Actually, what's going on below is a temporary connection (blue cable) from the control unit directly to the RCS thermostat just to verify communication between the two. Also, the RS485 loop is (for now) terminating at the control unit; therefore, you can see that the cable which will later continue to the upstairs furnace is disconnected. RCS actually recommends installing a 100 ohm resistor across the T+ and R- connection, but it shouldn't make any difference if your RS485 run is less than 100'.

The other thing of note is that of the four wires on the RS485 connection, only three are being used. The far left one (12+) is disconnected, and I'm using a transformer to power the CU instead. I'll have to be careful to ensure that this wire not become shorted. For me, it's delivering power to the keypad.

If you have HVAC control units in the RS485 run between Stargate and any keypads, you'll need to splice the 12+ wire together to make sure you have continuity to other devices (keypads?) that need power.

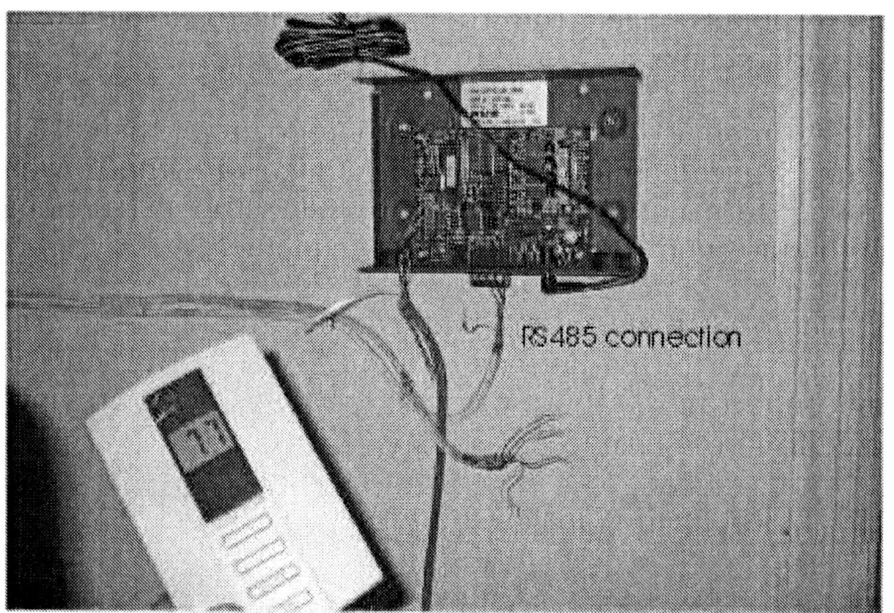

a temporary connection to verify that it works

Once communication has been verified, it's time to do the REAL connections between the thermostat, furnace, and control unit. Since I knew that this was going to be a retro, I told my HVAC guy to put in "el cheapo" thermostat until I was ready to install the RCS equipment.

Since I already have a thermostat wire directly connecting the furnace to the "el cheapo" upstairs, I now have the task of getting the control unit wired in between the two. Here's how it's done:

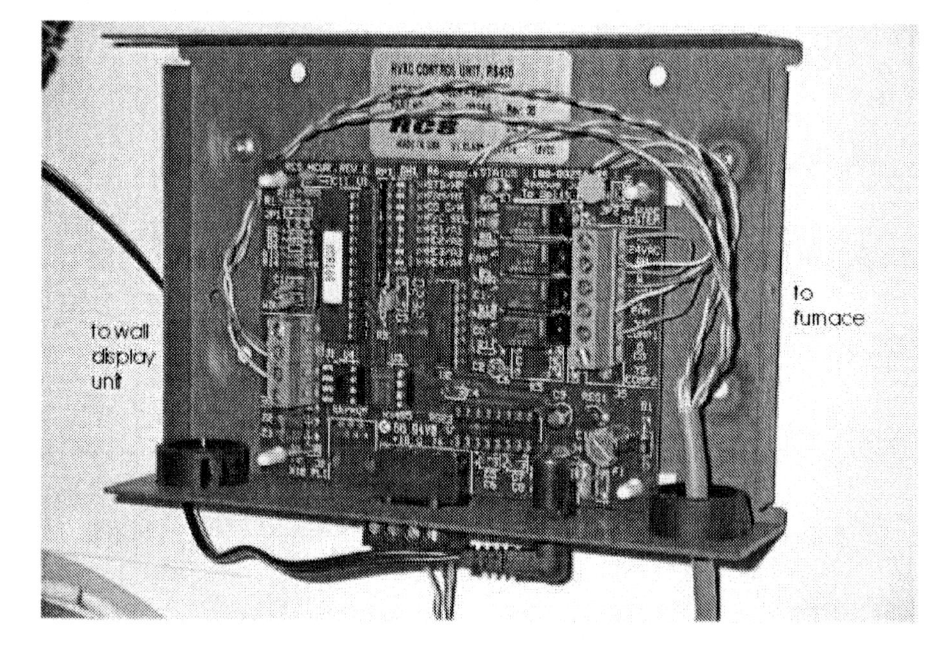

First of all, be sure to ***turn off the power*** to your furnace before you do this!

A single Cat5 cable is pulled from the control unit to where the thermostat wiring connects to the furnace.

In the illustration above, the green and brown pairs of the cat5 are wire-nutted to the furnace/compressor wiring. Which wires go to what will depend on your furnace; you just need to follow the instructions that come with the RCS unit (don't worry - it's explained).

On the left side of the control unit, you can see the blue and orange pairs which connect to the thermostat input. At the other end of the cable, they connect to the thermostat wiring (which was previously disconnected from the furnace).

Below is the mess of wiring inside the furnace.

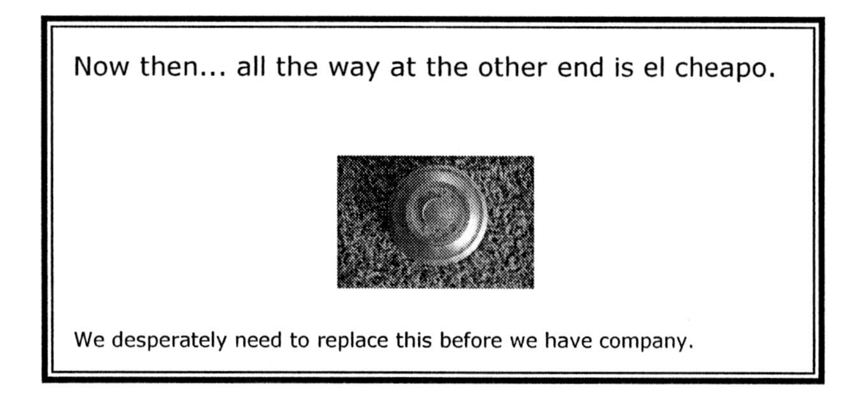

The old thermostat wasn't really mounted in the carpet. Seriously, it was on the wall, like the new one below (RCS calls it a "wall display unit"). It was simply a matter of connecting the existing wires to the "WDU", *making sure to use the terminals on the **right** side* (labeled "CU"). The terminal bank on the left side is for a remote temperature sensor.

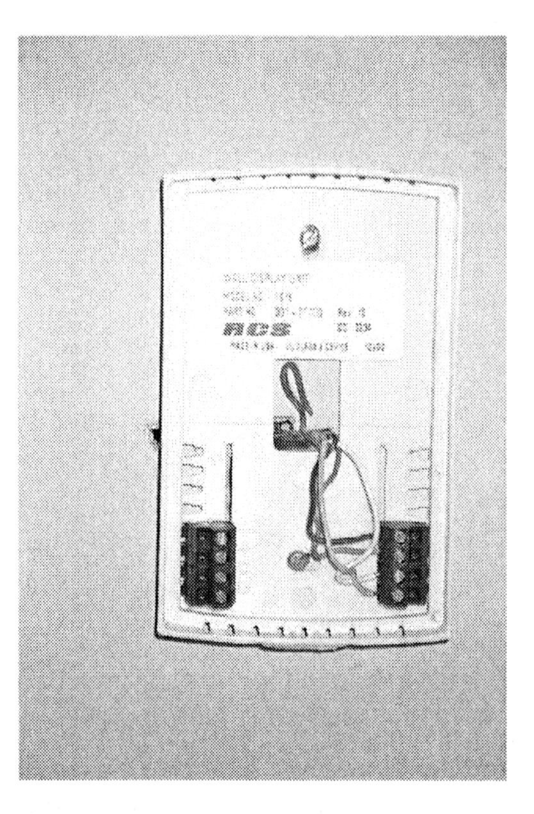

Double-check all your wiring! It's not too difficult to toast these things! The terminals here are marked "-", "+", "C", and "D". They correspond to "GRND", "V+", "C", and "D" on the control unit.

NOTE: RCS also makes an X10 version of the TR15 thermostat if you're not going to utilize an RS485 run. However, even though X10 *can be* fairly reliable, the less X10 devices you have running - the better! **Anytime you can hardwire, do so!**

A word about electronic dampers: I discussed these in volume one, but a quick review is probably appropriate in case you want to use them. While I already said that I don't really need any myself, I did run a couple of cat5's to the upstairs furnace ductwork. If ever I should change my mind, the wiring is in place.

In my case the damper(s) would be installed right above the furnace, or else in the attic. Just make sure that you choose a location that's accessible in the future (for service-related issues). It's simply a matter of inserting the damper into the ductwork and applying power via your cable run to the two terminals on the damper.

Your power source could be a 12v transformer back in the equipment room. The (normally open) damper will be controlled by running one of the two power leads through a Stargate relay.

Simply include events in your Stargate schedule to open/close the relay in order to control the damper. For instance, you could close the damper to an unused room at night & open it in the morning, etc. Or, if you're monitoring temperature in that room, your schedule can specify a setpoint at which you open or close the relay.

One **CAUTION:** closing ducts can potentially put your HVAC system under stress. Installing a barometric bypass can help alleviate this condition, but it might be a good idea to discuss what you're doing with a local HVAC technician & get his input.

The LCD96M Keypad:

There's nothing to the physical installation of this, especially if the RS485 run is in place.

Really the only thing for me to discuss might be programming the keypad. While there are some default menus already pre-programmed into the keypad, I had to modify some of them for my setup. Others I had little use for & simply created my own.

Most of the programming is done directly on the keypad from the WinEvm software. There are some added things you can do in your schedule, though, to enhance functionality.

Let's begin by looking at keypad programming (*I'm going to use my* ***house audio menu*** *as an example of what can be done*).

After I set the keypad's address and located the right keypad in WinEvm I was ready to go to work:

Figure 1

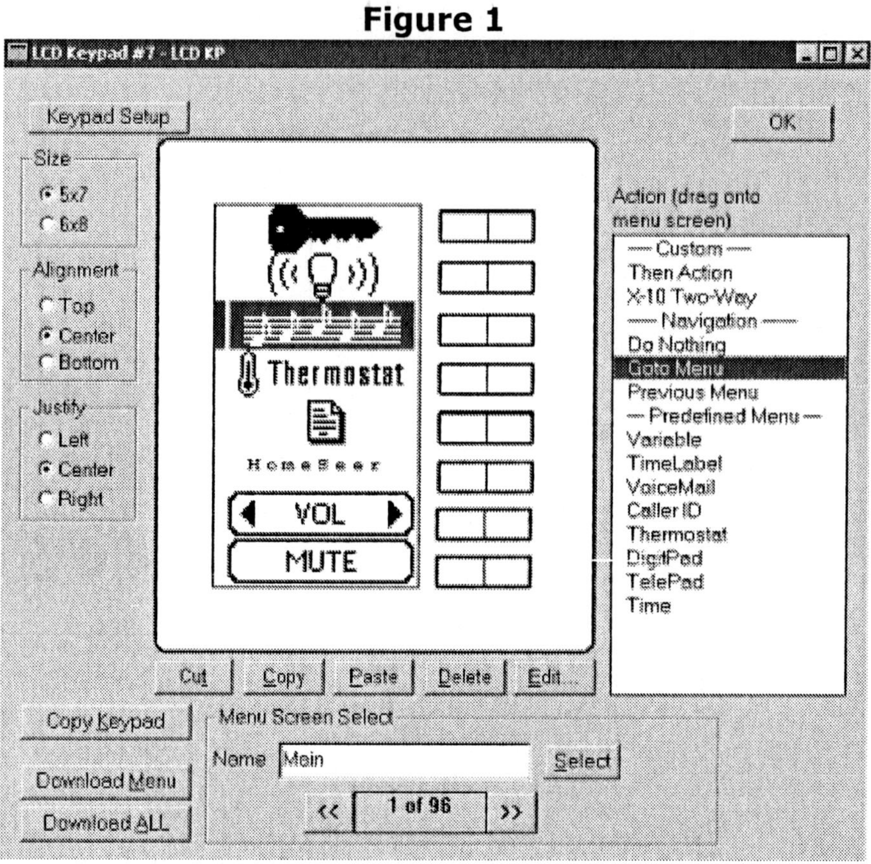

I used the "music" bitmap to represent my audio system.

Figure 2

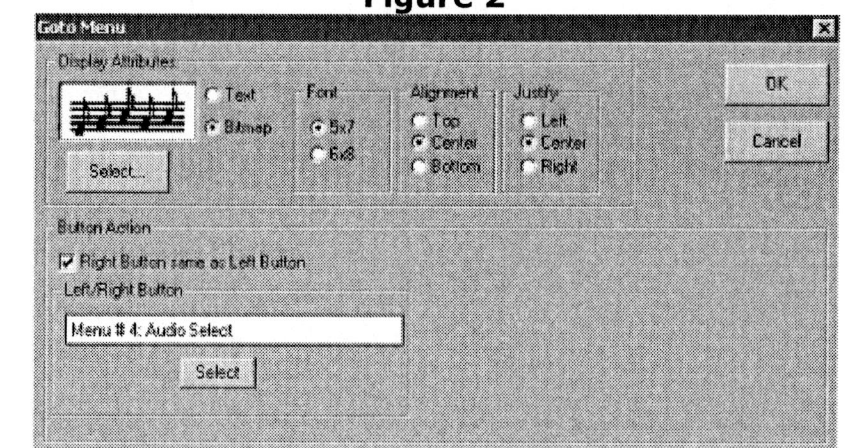

Figure 2 is the next screen that comes up during programming in which I designated menu #4 as the next menu to appear with the "music" buttonpress.

Figure 3

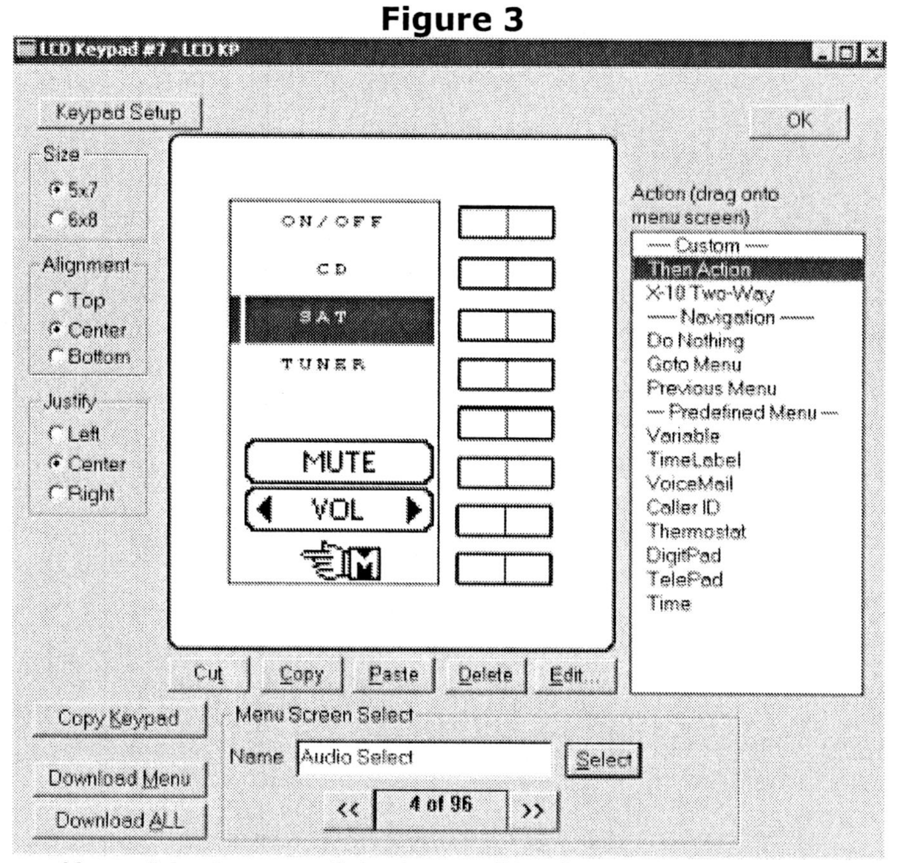

Menu #4. A press of the button "SAT" (satellite) results in a "Then Action," which could be an X10 or IR command or, in this case, a THEN MACRO (figure 4, below).

Figure 4

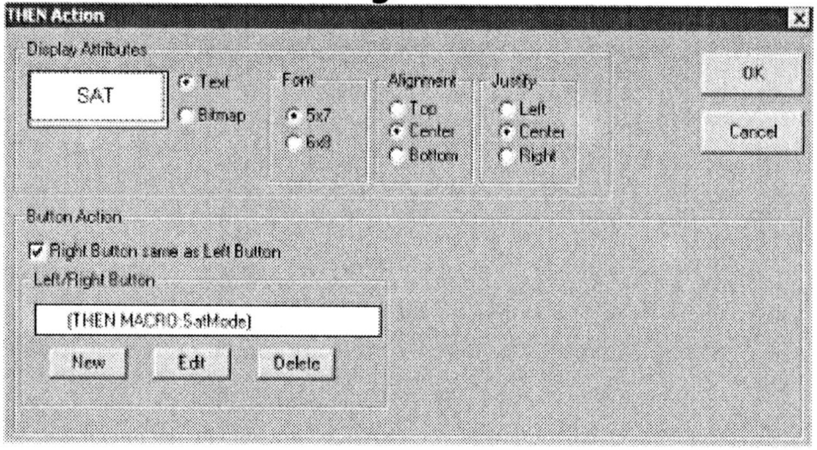

THEN Action
Display Attributes
SAT
Text
Bitmap
Font
5x7
6x8
Alignment
Top
Center
Bottom
Justify
Left
Center
Right
OK
Cancel
Button Action
Right Button same as Left Button
Left/Right Button
(THEN MACRO:SatMode)
New
Edit
Delete

Figure 5

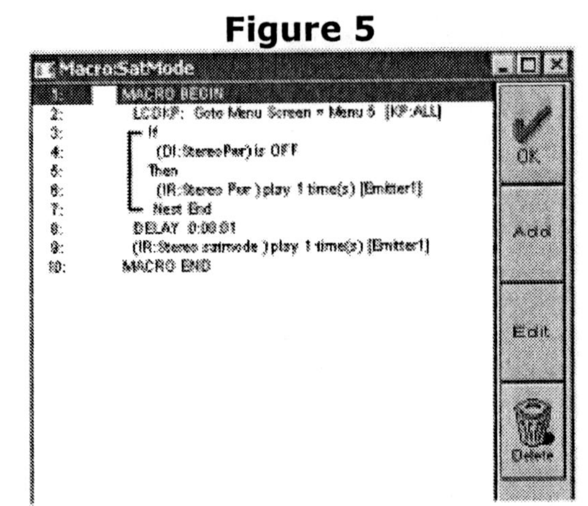

Here's then-macro "sat mode". It checks to see whether it needs to turn on the stereo, and then changes the receiver to the "satellite" mode. At the same time, the macro also sends me to the next menu where I'll choose my music channel (figure 6 below).

Figure 6

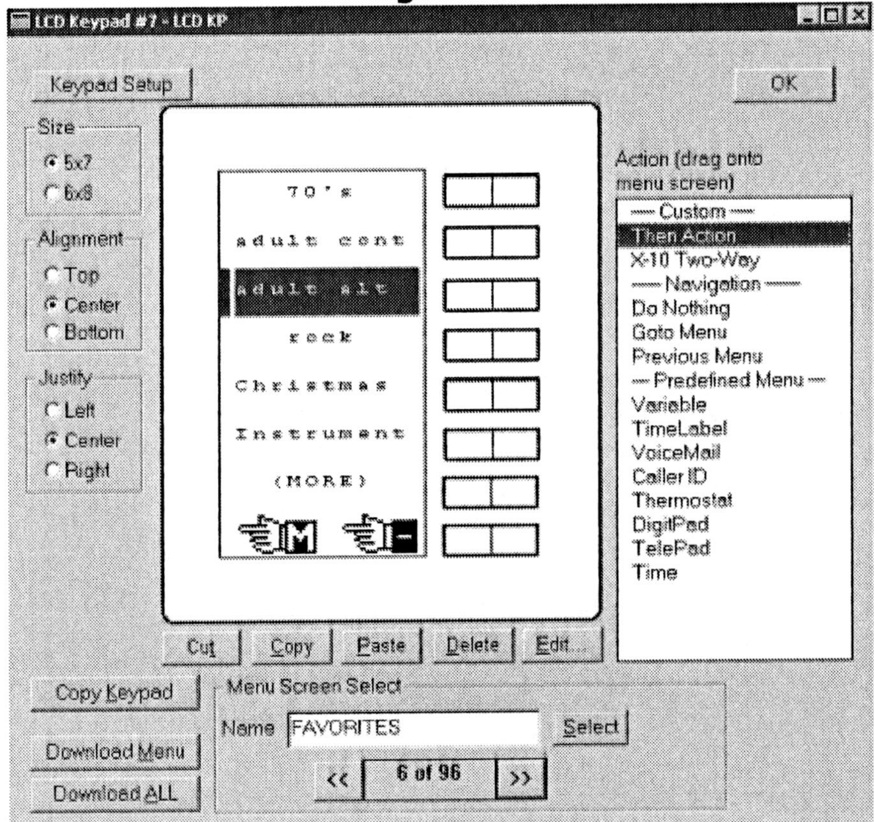

A buttonpress here executes another THEN MACRO.

Figure 7

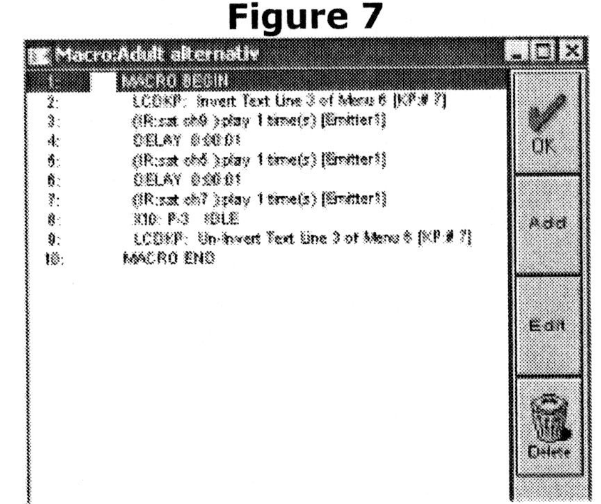

The Macro begins by inverting the selected line of text on the keypad (just to show that the command was received). It then issues the appropriate IR commands and finishes up by un-inverting the line of text.

Hmmm.... That X10 P3 IDLE command shouldn't be there in Fig. 7.... Oh well, must be left over from some vain experiment.

Anyway, you can see how there are abundant things you can do by combining the schedule with the keypad programming. Obviously I've done a lot more with my LCD96M than I've shown here, but there's no way (not to mention no real need) to show everything.

But here's a list of things that I either have done, or will do eventually, with the keypad:

- ***Centralized HVAC control***
- ***Audio control***
- ***Security system (as an additional security keypad)***
- ***Lighting control, by scenes or individual lights***
- ***Message board, displaying time/temp/or notes***
- ***3rd party add-ons, like Homeseer.***

Miscellaneous:

- Garage Door
- Fishing a Wall
- Gas Fireplace
- X10 and Table Lamps
- Homeseer

Garage door:

I once wrote a newsletter dealing with this from an X10 standpoint. Though I personally didn't utilize X10 for *my* garage door control, I've included the article here for your consideration:

Automating your Garage Door

I recently moved out into the country and am having to learn a whole new way of life. For instance, I'm learning an entirely new sense of dread when it comes to garbage night since I now have a 400' long driveway.

I'm also discovering what it's like to have numerous critters making themselves at home in my backyard -

- and in my garage.

One morning I woke up especially early to hear something *(correction - I mean some THING)* clawing at the door which leads to my garage. To make a long story short, I had left the garage door open overnight and a raccoon was all over my stuff. It even pottied on top of my 2002 Trans Am.

Here's how to make sure that doesn't happen to you (or me - again)!

Have that garage door close by itself at night. At least have it give you some kind of reminder that it's open. Let me give you a couple of ways to deal with this problem:

The simplest way is to install a wireless garage door sentry. At about 50 bucks, it's quick and easy. This little thingie beeps anytime the garage door opens or closes; and it emits a green light when the door is closed, and flashes red when it's open. It won't automatically close the door for you, but it will at least give you some kind of alert.

The second way is a little more involved, but will give you significantly more control. You'll need an X10 Universal Module, an X10 Powerflash Module, and some kind of intelligent X10 controller (i.e. PC based software such as Homeseer or a JDS Product).

First, you need to understand that when you press the button to open or close your garage door, you're simply operating a momentary **contact-closure** that completes a circuit. The voltage triggers your opener to do its thing. The Universal Module does the same thing, but is based on X10 signals rather than your finger presses.

You'll need to connect a 2-conductor wire **in parallel** with your garage door opener button. Probably the neatest place to splice into the existing wiring is right at the opener (I'd suggest you **unplug your opener first**). The other end of the 2-conductor connects to the terminals on the Universal Module. Set the X10 address on the module (for this illustration **B-1**) and you can open/close the garage door via X10!

Oh - and make sure you set the Universal Module for **MOMENTARY** contact-closure!

Now you need a way to know whether your door is open or closed. Here's where the Powerflash Module comes into play. You first of all need a normally-closed security sensor on your garage door. If you already have a security system that monitors your garage door, you're set. Just connect the Powerflash to the sensor wiring (in series).

Otherwise, you'll need to install the sensor and run the wiring to the terminals on the Powerflash. Put the Powerflash in mode 3, input B, and set its address to something other than the Universal Module (let's say **D-1**).

Now you need to program your controller (it needs to be capable of IF/THEN logic). Let's say you want your garage door to automatically close at dark. It would read something like this:

IF it's dark
  AND
IF D-1 OFF *(garage door open)*
  THEN B-1 ON *(momentary contact-closure to garage door)*

Now your garage door is set to automatically close every night. No more raccoon messes on your car!

IMPORTANT! *Most newer garage doors have built-in safety features like IR beams to prevent closing doors on people/objects. Please make sure you use* *common sense in this application. Garage doors that automatically close have the potential of doing property damage or bodily harm.*

Well, that's one way of doing it. However, always better to hardwire - so this is what I did:

Heck, now that I think about it, this doesn't need much explaining beyond what's said in the newsletter above. The only difference is that the opener wiring runs all the way back to Stargate's relays (instead of a Universal Module), and the security sensor runs through Digital inputs for status detection. The X10 method just means you shouldn't have to run wire beyond the nearest electrical outlet.

garage door opener button and 2-conductor

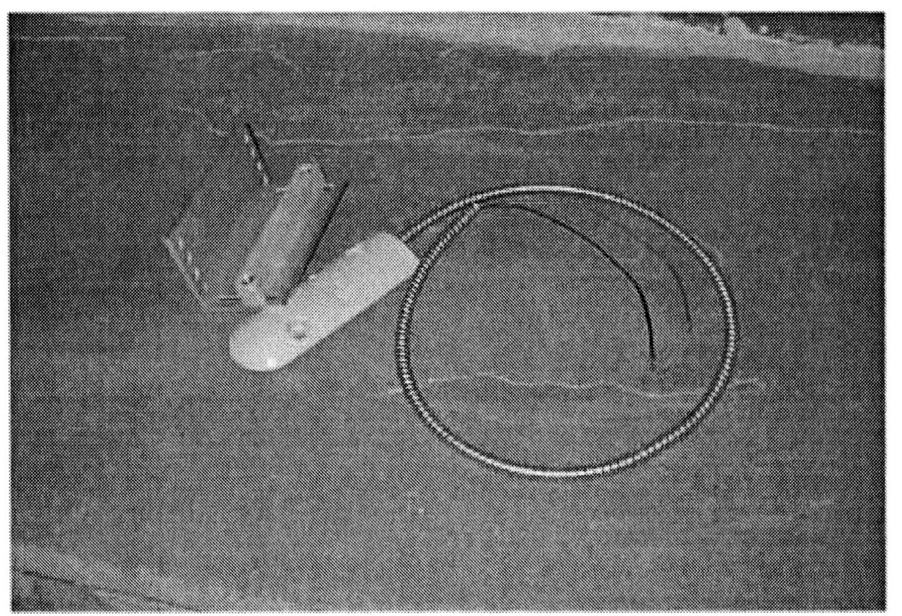

garage door sensor

Once I connect the 2-conductor (above left) to the button's terminals, I'll be done. Then it'll be just a matter of programming.

Fishing wall:

Nuts. Sometimes even if you thought you'd run everything you might need, you (that is, I) find that you forgot something.

Well, I forgot to run the wiring for the security system keypad. So - I'm afforded the opportunity to do a little retro work for you. This wasn't too difficult since it was in the basement & the ceiling was unfinished. This allowed me to drill into the plate above the wall and push my fishtape down through the insulation to the cutout I'd made for the keypad.

Then I just had to tie a string onto the end of the fishtape and pull it back up. Next, I reversed the process by taping my cable to the string and pulling it back down.

You can obviously do this same procedure from your attic, but if you don't have access to the area immediately above the room, and if you also have a drywalled ceiling, your life gets difficult.

In such a case you'll likely be cutting holes in the wall AND ceiling to get wire where it needs to be (unless you can do it from an unfinished basement).

But drywall is made for patching. Just hope you don't have wallpaper....

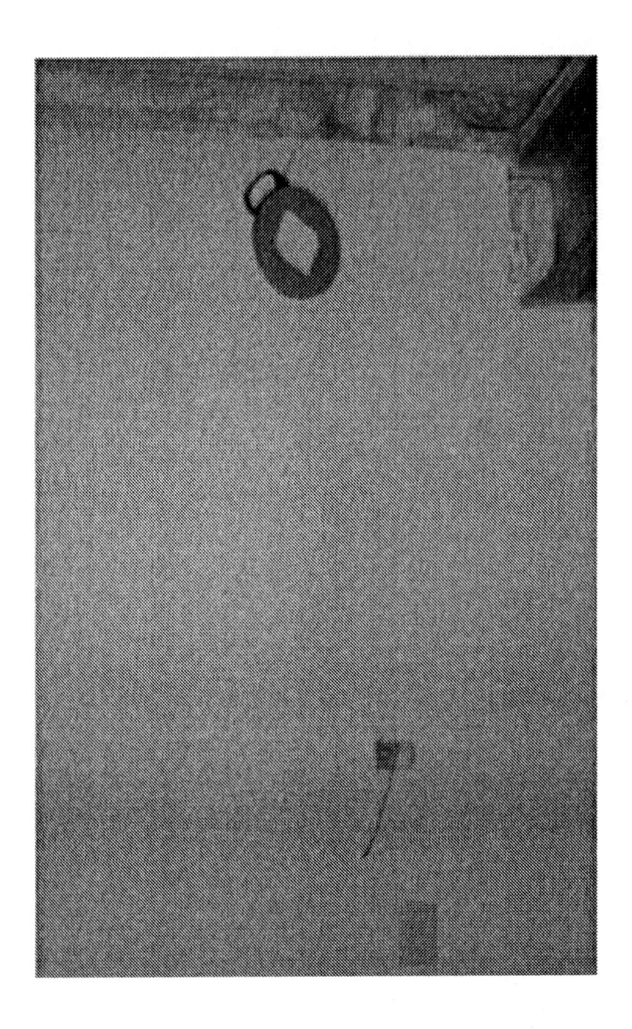

Gas Fireplace:

I wanted the ability to automatically turn the fireplace on or off . Usually a gas fireplace can be controlled by a low-voltage switch which has a single 2-conductor running to the fireplace. The switch is just a contact-closure item (e.g. a plain ol' electrical switch).

However, I needed to accommodate certain safety concerns, namely the need to manually override everything really fast. After all, we're talking fire here.

There were a couple of ways to do this, depending on how much control I wanted. In either case, I needed to run a 2-conductor from a set of Stargate relays to the gas logs.

The first way is to install a single electrical switch somewhere in the run (typically close to the fireplace). I'd simply pass one of the two conductors through the switch relays as the below diagram shows. If I keep that switch closed, Stargate automation can start/stop the fireplace; and in an emergency I can flip the switch to stop the fireplace manually.

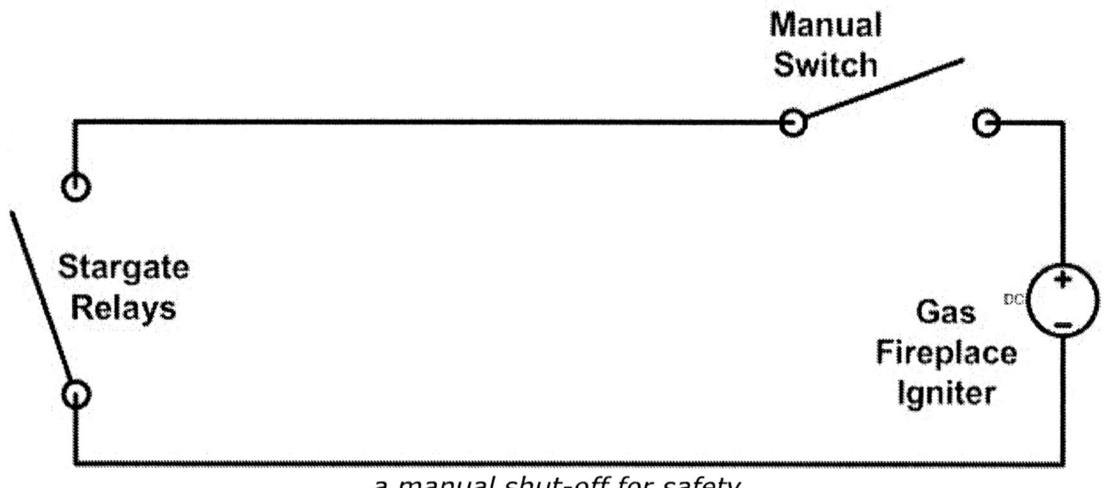

a manual shut-off for safety

The one problem with this design is that I cannot START the fireplace manually. Therefore, with a little thought I came up with the following scenario:

I simply installed a second switch right next to the first in parallel with the Stargate relays. If switch "A" is left in the closed position, the entire setup becomes a manual operation via switch "B." If I leave switch "A" open, I have the same setup as in the above diagram. I just have to keep track of *which switch is which* (*oooh! - try saying "which switch is which" 4 times fast*).

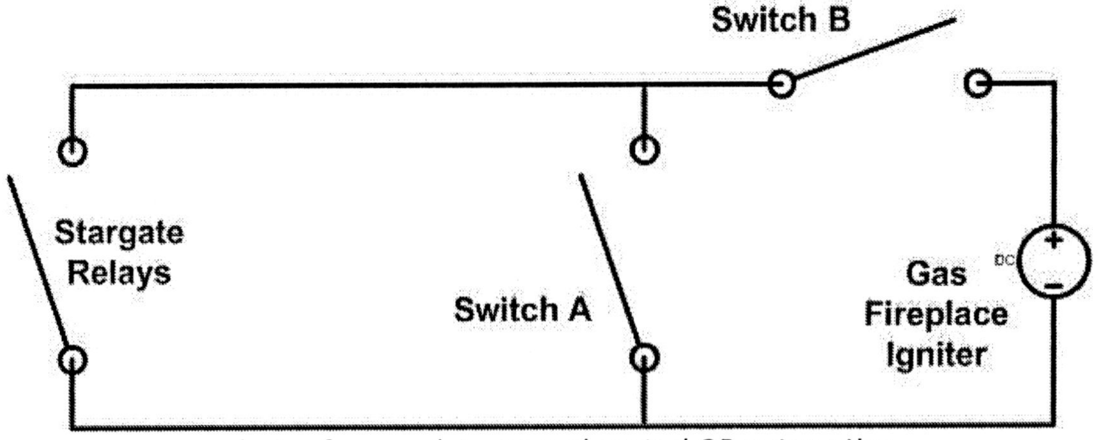

a design for complete manual control OR automation

NOTE: *You should use good sense when doing something like this. For instance, if you have pets or children that like to play in the fireplace, you might want to think twice about automating this feature.*

Controlling Table Lamps with X10

Since lamps don't always have a switch (unless they're on switched outlets), they need to be handled a little differently than other lights. The most elegant solution is to install an X10 outlet. You can typically find such things at SmarthomeUSA.com or other distributors at reasonable prices.

Often these units contain one outlet that's X10-based, and one that isn't. After all, there are certain things in your home that you don't want controlled by X10 (like your computer, etc.).

If you don't to bother with changing outlets, you can also use a wireless X10 base with remote and a normal X10 module. Simply plug your lamp into the module, the module into an outlet, and the base into another outlet. When you set the base, remote, and module to a common address you can use the remote (or Stargate's schedule) to turn your lamp(s) on/off.

It's even possible to manually override the X10 module. Typically, you can turn the lamp on at the lamp switch by turning it on, off, then on again; and of course you can turn it off manually.

The only thing to bear in mind is that X10 can only control the light if the light switch on the lamp itself is left in the "on" position.

Homeseer (a third party product)

Here's a quick summary of what Homeseer is/does (from newsletter February 2003):

> ... Homeseer is a software based application that has some powerful features. At around 100 bucks (plus your X10 interface), it's definitely affordable for most of us.
>
> It mainly utilizes X10 to communicate, but there are numerous plug-ins available which make it compatible with other HA devices (including Stargate). Naturally, it has a Windows interface for easy programming of X10 scheduling, timers, etc., but the real power of Homeseer is in its ability to utilize VBscript or Javascript (though its native environment is VBscript).
>
> Even for those of us who aren't comfortable with scripting, the documentation is quite good. And I have to compliment the makers of Homeseer for their willingness to allow the user community to develop and share scripts, plug-ins, etc. - right on Homeseer's website!
>
> You can literally download a script, copy and paste it into an event, and (barring any bugs) it works. If you happen to have an always-on internet connection, there's some fairly neat stuff you can do. For instance, one script (which I believe actually comes with Homeseer) retrieves the local weather forecast and plays it audibly over the PC.
>
> There are lots of other neat scripts available. People are always coming up with new ideas & are usually eager to share!
>
> But let me tell you about what I think is one of Homeseer's outstanding features: Homeseer has a fairly impressive voice-recognition engine that's actually reliable. When I tested it using a standard cheap PC microphone, it rarely missed. What's more, using Microsoft Agent and text-to-speech, Homeseer can talk back to you in a - hmm - well, *almost* natural sounding voice ....

The only unfortunate thing about Homeseer is that the software has to be running at all times on my basement PC in order to perform. Not that I mind (my computers run 24/7 anyway) - but it *is* one more thing that I have to keep an eye on.

One much desired feature that's native to Homeseer is a **Web Interface**. With this you can access your system from any PC in your home, or even over the Internet (if you're behind a firewall you'd have to open a port. Your firewall's documentation should instruct you). JDS also has a **WebXpander** for much the same purposes. I'll talk about that in a moment.

So far, I've just begun to tap into Homeseer's resources. Let me share with you the "weather" thing that's mentioned above, & how I implemented it.

Here are a couple of screen shots showing the simple configuration in Homeseer itself. I simply configured the software to run the script when it detects the "C7" X10 signal.

Event Properties (get weather)
Infrared Control | Dial-up Connection
Trigger | Device Actions | App/Sound/Email | Scripts/Speech | Run Events
Description
Event Name:
get weather
Type: By X10 Command
Group: system
Voice Command
get weather
Enable: Disabled
Edit
Confirm
Trigger if the given X10 housecode and unitcode are received
House C
Unit 7
Command: On
Attributes
Disable this trigger
Delete after trigger
Apply Conditions to trigger
Web page guests can trigger
Do not log this event
Include in power fail recovery
Do not re-trigger within minute
Days
Choose Days
Every Day
Mon Tue Wed Thur Fri
Sat Sun
OK Cancel Help

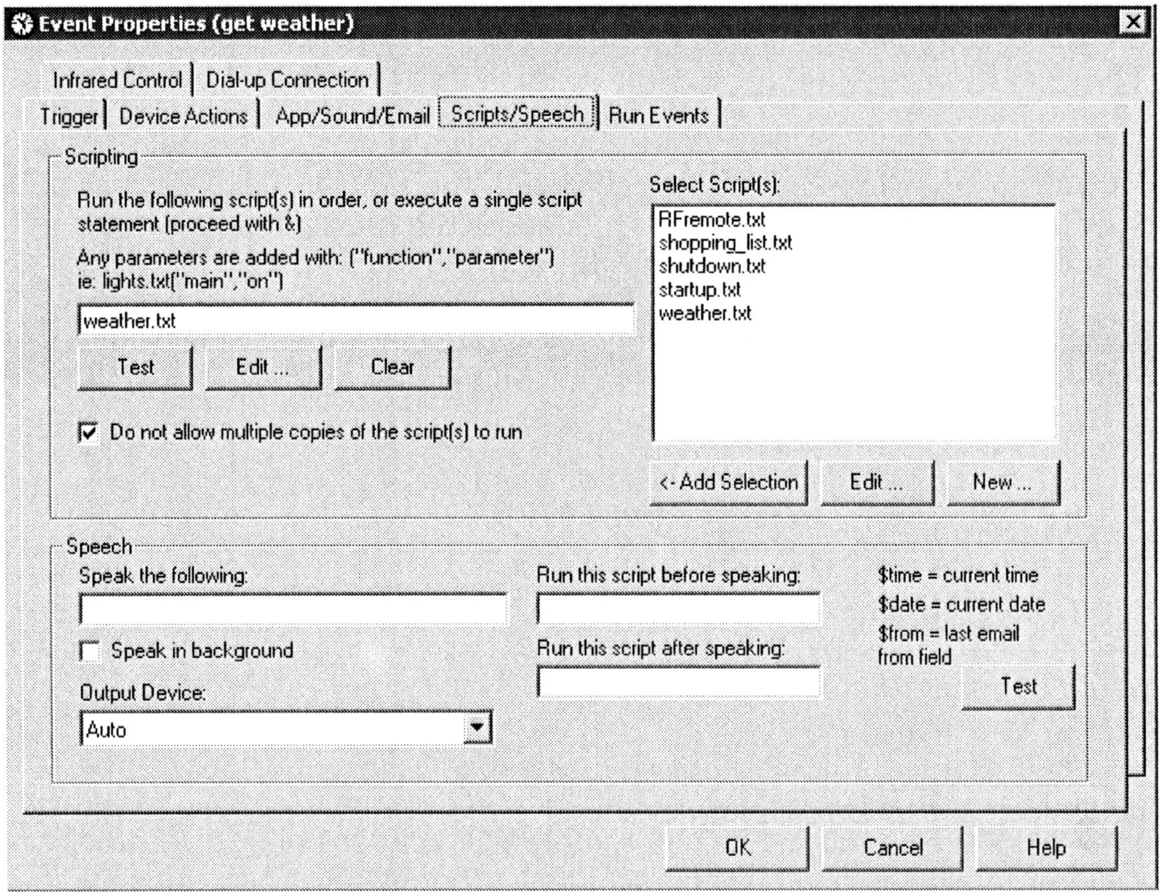

Event Properties (get weather)

Infrared Control | Dial-up Connection
Trigger | Device Actions | App/Sound/Email | Scripts/Speech | Run Events

Scripting

Run the following script(s) in order, or execute a single script statement (proceed with &)

Any parameters are added with: ("function","parameter") ie: lights.txt("main","on")

weather.txt

Test | Edit ... | Clear

Do not allow multiple copies of the script(s) to run

Select Script(s):
RFremote.txt
shopping_list.txt
shutdown.txt
startup.txt
weather.txt

<- Add Selection | Edit ... | New ...

Speech

Speak the following:

Speak in background

Output Device:
Auto

Run this script before speaking:

Run this script after speaking:

$time = current time
$date = current date
$from = last email from field

Test

OK | Cancel | Help

I've designed a "Homeseer" button on my Stargate keypad. This will lead me to a page where I (will) have various Homeseer activities. In this case, a "weather" button will execute the X10 command "C7 ON" which launches the Homeseer weather script.

In order to play it over my house audio system, I have the following lines in my Stargate schedule:

```
EVENT: Get Weather
If
  X10:C-7 ON
Then
  Audio:Connect Line Level In to Spkr
  THEN MACRO:SG spkr switch)
  DELAY 0:00:27
  Audio:Disconnect Line Level In to Spkr
  (THEN MACRO:Ster spkr switch)
  X10:C-7 OFF
End
```

A buttonpress on the keypad executes "C-7 ON." Stargate routes line-level audio from my PC sound card to Stargate speaker output, does the relay speaker-switch routine I discussed in the House Audio section; and then it delays long enough (27 seconds) for the message to be delivered before setting everything back to where it originally was.

Soon I'll be adding the microphone in the kitchen. At that point I'll be able to just ask for the weather. Can't wait for more toys!

JDS WebXpander

This is a new product and unfortunately for me, I don't even have one yet. But enough of us want network/internet access to our homes that it would be remiss of me not to tell you about it.

It's been kind of an issue that you could only access Stargate from a single PC via a serial connection until now. With the WebXpander, any browser can take you to your Stargate for direct control of the home, or even for modifying and downloading new schedules. It also adds a degree of email functionality much as Homeseer does. To read more, check out the JDS site at www.jdstechnologies.com.

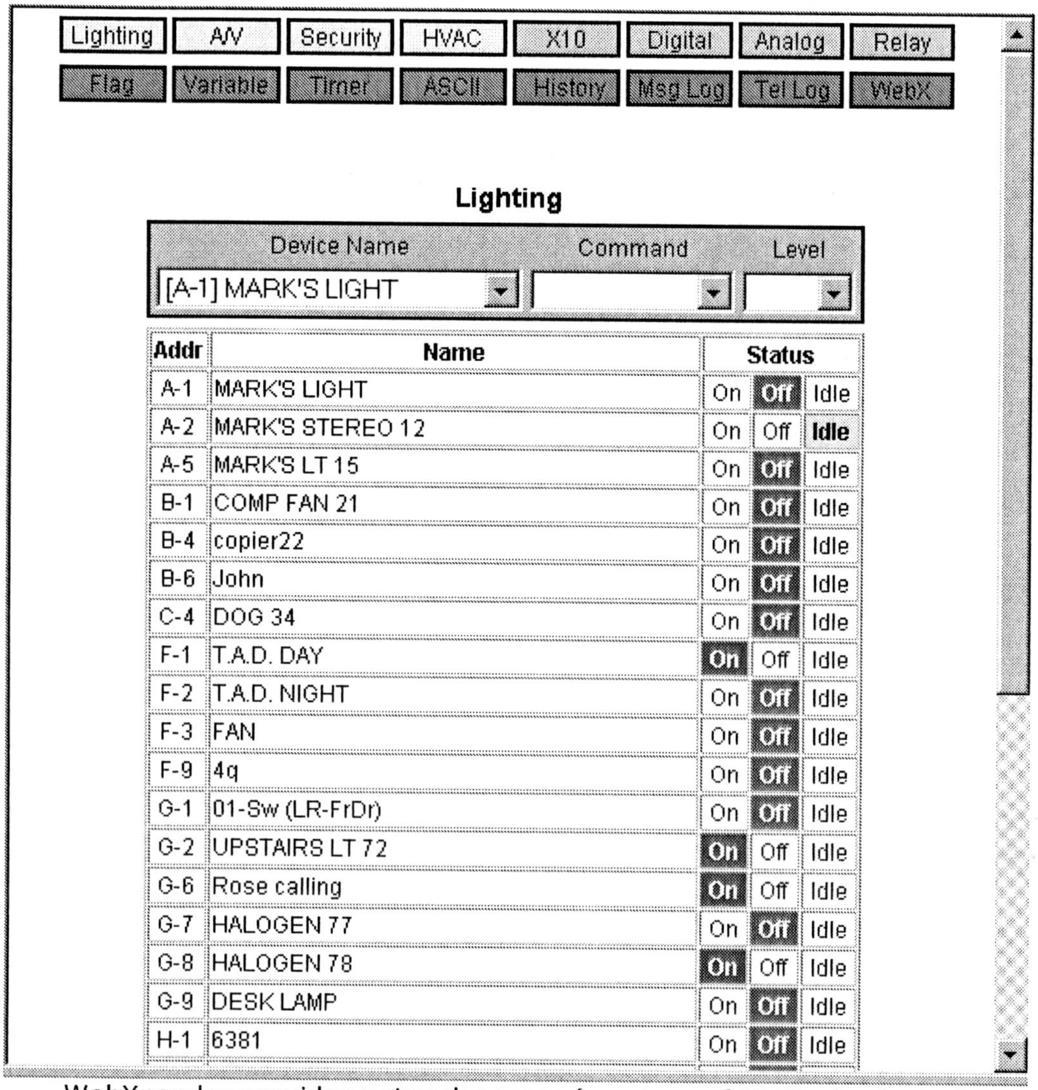

Lighting	A/V	Security	HVAC	X10	Digital	Analog	Relay
Flag	Variable	Timer	ASCII	History	Msg Log	Tel Log	WebX

Lighting

Device Name	Command	Level
[A-1] MARK'S LIGHT		

Addr	Name	Status		
A-1	MARK'S LIGHT	On	Off	Idle
A-2	MARK'S STEREO 12	On	Off	Idle
A-5	MARK'S LT 15	On	Off	Idle
B-1	COMP FAN 21	On	Off	Idle
B-4	copier22	On	Off	Idle
B-6	John	On	Off	Idle
C-4	DOG 34	On	Off	Idle
F-1	T.A.D. DAY	On	Off	Idle
F-2	T.A.D. NIGHT	On	Off	Idle
F-3	FAN	On	Off	Idle
F-9	4q	On	Off	Idle
G-1	01-Sw (LR-FrDr)	On	Off	Idle
G-2	UPSTAIRS LT 72	On	Off	Idle
G-6	Rose calling	On	Off	Idle
G-7	HALOGEN 77	On	Off	Idle
G-8	HALOGEN 78	On	Off	Idle
G-9	DESK LAMP	On	Off	Idle
H-1	6381	On	Off	Idle

WebXpander provides network access (courtesy of JDSTechnologies)

Summary:

As I write this, everything is in a relative state of workingness. I know that's not good English, but I'd like to keep a sense of humor about some of the stuff I'm about to tell you.

I still have more to do in regard to tweaking my schedule. I've kept the few samples in this book fairly simple *(with the exception of the theater room sample on the last page)* so that things will be understandable.

I surely hope that you've found my work a little enlightening, and that you've found enough examples to inspire you in your own "adventure."

The last thing I'll share with you is a list of things I went through, because nothing is ever as easy as a book makes it look.

Once all the hardware is hooked up it's tempting to think "Whew - it's a relief to have all that stuff in place. Now all I have to do is program."

In a perfect world the rest *should* have been cake. But in truth it wasn't.

If you'd like to hear me moan in specifics you can check out my section on "headaches" on the next page. Otherwise, I've now cleared my conscience.

In summary I have the following systems connected in some fashion to Stargate:

- *telephones*
- *pc's*
- *X10 network*
- *IR system*
- *audio*
- *security*
- *cameras*
- *HVAC*
- *keypads*
- *garage door*
- *driveway sensor*
- *fireplace*

This is far from exhaustive. Most of us who've done their own systems will agree that these things evolve over time, which in fact, is the big advantage of doing it myself. My needs will change, my interests and lifestyle will also - and two years from now the list might be bigger - or maybe just different.

The point is that it's never really done.

Oh man, I seem to be in the mood to pontificate! I'm supposed to be talking about nuts 'n bolts here, right? OK.... sorry.

In the earlier sections I shared some events that I had written to control HVAC, audio, & other compartmentalized bits of the whole. For the most part, it's about that easy.

But there are times when one event conflicts with another, or when you might wish for one system's behavior to be contingent upon another.

For example:

> *"I'd like the TV to automatically switch to the front door camera at night when motion is detected, but only when the security system is armed in "stay at home" mode;*
>
> *If, however, the security system is armed and nobody is home, I'd like the VCR to record the visitor's presence - unless of course, the driveway sensor did NOT sense a vehicle - in which case my webcams will simply monitor the interior of the house - and Stargate will call my cell phone with a message about a visitor;*
>
> *...unless it's the third Thursday of August, when I simply MUST have a completely different schedule of events to run!"*

Things can get pretty weird. Actually, that example is nothing next to some of the things that have to be done in a Theater Room (depending on complexity and number of components).

In order to demonstrate what I mean (and also since mine is relatively simple), I've included some screenshots from a schedule I wrote for a customer of mine.

It's an old script and I can tell that for some reason a few things aren't quite right (missing IR commands, etc.), but it's meant to be merely an example of how lengthy things can get. You'll have to study this to see the relationships between components, cameras, flags, etc....

This is a theater room with a THX Marantz system and overhead DLP projector. Three outdoor cameras are modulated through the VCR. *Please ignore the red highlighted text - WinEVM does this when the items (the customer's) in the schedule aren't present in the (my) device database.*

I gained control over things through the use of a number of flags which, when set, launched certain events. To the left (below) is most of the activity concerning the theater room itself. On the right are the "fast events" controlling the security cameras.

```
714:  EVENT: System Off
715:    If
716:      IR Seq:' ' Received within 5 seconds
717:        -OR-
718:      (DI:projr probe) Goes OFF
719:    Then
720:      (F:system on) IDLE
721:    End
722:
723:  EVENT: System Off 2
724:    If
725:      (F:system on) is IDLE
726:    Then
727:      (F:DVD Play) CLEAR
728:      (F:VidPlay) CLEAR
729:      (F:front dr camera) CLEAR
730:      (F:back camera) CLEAR
731:      (F:side camera) CLEAR
732:      X10: B-11 PRE-Set Level 26 %
733:      (IR:ProjectorOff ) play 1 time(s) [Emitter1]
734:      (IR:DvdStop ) play 1 time(s) [Emitter1]
735:      (IR:VidStop ) play 1 time(s) [Emitter1]
736:      DELAY 0:00:02
737:      (IR:sr18Off ) play 1 time(s) [Emitter1]
738:      (F:system on) CLEAR
739:      (F:PowerOn) CLEAR
740:      DELAY 0:00:10
741:      X10: B-11 PRE-Set Level 90 %
742:      (IR:sr18Off ) play 1 time(s) [Emitter1]
743:    End
744:
745:  EVENT: Power On
746:    If
747:      (F:PowerOn) is SET
748:        -OR-
749:      (DI:projr probe) Goes ON
750:    Then
751:      (F:PowerOn) SET
752:      (F:system on) SET
753:      (IR:ProjectorOn ) play 1 time(s) [Emitter1]
754:      (IR:sr18Power ) play 1 time(s) [Emitter1]
755:      DELAY 0:00:11
756:      X10: B-11 PRE-Set Level 6 %
757:      DELAY 0:05:00
758:      If
759:        (F:system on) is SET
760:      Then
761:        X10: B-11 OFF
762:      Nest End
763:    End
```

```
271:  FASTEVENT: camera view/reco [ X10:D-2 Goes ON]
272:    Then
273:      (THEN MACRO:cam-bk2 motion)
274:      DELAY 0:00:03
275:      If
276:        (F:system on) is SET
277:        and Time is After (TL:dark)
278:      Then
279:        (F:VidPlay) SET
280:        DELAY 0:00:02
281:        (F:back camera) SET
282:      Nest End
283:    End
284:
285:  FASTEVENT: camera view/reco [ DI: Goes ON]
286:    Then
287:      (THEN MACRO:cam-side2 dbell)
288:      Voice:SIDE DOORBELL [Spkr]
289:      DELAY 0:00:03
290:      If
291:        (F:system on) is SET
292:      Then
293:        (F:VidPlay) SET
294:        DELAY 0:00:02
295:        (F:side camera) SET
296:      Nest End
297:    End
298:
299:  FASTEVENT: camera view/reco [ DI: Goes ON]
300:    Then
301:      (THEN MACRO:cam-front2 dbell)
302:      DELAY 0:00:03
303:      If
304:        (F:system on) is SET
305:      Then
306:        (F:VidPlay) SET
307:        DELAY 0:00:02
308:        (F:front dr camera) SET
309:      Nest End
310:    End
```

```
764:
765:  EVENT:  System on/ DVD Play
766:      If
767:       (F:system on) is CLEAR
768:       and  IR Seq:' ' Received within 3 seconds
769:      Then
770:       (F:PowerOn)  SET
771:       (F:system on)  SET
772:       (F:DVD Play)  SET
773:      End
774:
775:  EVENT:  DVD Play
776:      If
777:       (F:system on) is SET
778:       and  IR Seq:' ' Received within 3 seconds
779:      Then
780:       (F:DVD Play)  SET
781:      End
782:
783:  EVENT:  DVD Flag
784:      If
785:       (F:DVD Play) is SET
786:      Then
787:       (IR:sr18Power ) play 1 time(s) [Emitter1]
788:       (F:VidPlay)  IDLE
789:       (F:front dr camera)  IDLE
790:       (F:back camera)  IDLE
791:       (F:side camera)  IDLE
792:       (IR:ProjComponent ) play 1 time(s) [Emitter1]
793:       (IR:VidStop ) play 1 time(s) [Emitter1]
794:       (IR:sr18DVD ) play 1 time(s) [Emitter1]
795:       (IR:DvdPlay ) play 1 time(s) [Emitter1]
796:        DELAY  0:00:03
797:       (IR:sr18DVD ) play 1 time(s) [Emitter1]
798:       (IR:ProjComponent ) play 1 time(s) [Emitter1]
799:       (F:DVD Play)  IDLE
800:      End
801:
802:  EVENT:  System on/ Vid Play
803:      If
804:       (F:system on) is CLEAR
805:       and  IR Seq:' ' Received within 3 seconds
806:      Then
807:       (F:PowerOn)  SET
808:       (F:system on)  SET
809:       (F:VidPlay)  SET
810:      End
```

```
311:
312:  FASTEVENT:  camera view/reco [ DI: Goes OFF]
313:      Then
314:       (F:front motion)  SET
315:       If
316:         Time is After (TL:dark)
317:        Then
318:         X10: B-4   PRE-Set Level  100%
319:         DELAY  0:01:00 Re-Triggerable
320:         X10: B-4   OFF
321:        Nest End
322:       If
323:         Time is After (TL:dark)
324:         and  (F:goodnight) is Not IDLE
325:        Then
326:         Voice:MOTION AT FRONT DOOR  [Spkr,ICM]
327:        Nest End
328:       If
329:         (F:system on) is SET
330:         and  Time is After (TL:dark)
331:        Then
332:         (F:VidPlay)  SET
333:         DELAY  0:00:02
334:         (F:front dr camera)  SET
335:        Nest End
336:       DELAY  0:00:20
337:       (F:front motion)  CLEAR
338:      End
```

```
811:
812:   EVENT: Vid Play
813:     If
814:       (F:system on) is SET
815:       and IR Seq:' ' Received within 3 seconds
816:     Then
817:       (F:VidPlay) SET
818:     End
819:
820:   EVENT: Vid Flag / cameras
821:     If
822:       (F:VidPlay) is SET
823:     Then
824:       (IR:sr18Power ) play 1 time(s) [Emitter1]
825:       (F:DVD Play) IDLE
826:       (F:front dr camera) IDLE
827:       (F:back camera) IDLE
828:       (F:side camera) IDLE
829:       (IR:DvdStop ) play 1 time(s) [Emitter1]
830:       (IR:ProjComposite ) play 1 time(s) [Emitter1]
831:       (IR:sr18VCR ) play 1 time(s) [Emitter1]
832:       DELAY 0:00:03
833:       (IR:ProjComposite ) play 1 time(s) [Emitter1]
834:       (F:VidPlay) IDLE
835:     End
836:
837:   EVENT: trigg backcam
838:     If
839:       IR Seq:' ' Received within 3 seconds
840:     Then
841:       (F:VidPlay) SET
842:       DELAY 0:00:01
843:       (F:back camera) SET
844:     End
845:
846:   EVENT: trigg frontcam
847:     If
848:       IR Seq:' ' Received within 3 seconds
849:     Then
850:       (F:VidPlay) SET
851:       DELAY 0:00:01
852:       (F:front dr camera) SET
853:     End
854:
855:   EVENT: trigg sidecam
856:     If
857:       IR Seq:' ' Received within 3 seconds
858:     Then
859:       (F:VidPlay) SET
860:       DELAY 0:00:01
861:       (F:side camera) SET
862:     End
```

```
863:
864:    EVENT: back camera
865:      If
866:        (F:back camera) is SET
867:      Then
868:        (F:front dr camera) IDLE
869:        (F:side camera) IDLE
870:        (IR:VidStop ) play 1 time(s) [Emitter1]
871:        (IR:Vid 1 ) play 1 time(s) [Emitter1]
872:        (IR:Vid 2 ) play 1 time(s) [Emitter1]
873:        (IR:Vid 3 ) play 1 time(s) [Emitter1]
874:        (IR:VidEnter ) play 1 time(s) [Emitter1]
875:        (F:back camera) IDLE
876:      End
877:
878:    EVENT: front camera
879:      If
880:        (F:front dr camera) is SET
881:      Then
882:        (F:back camera) IDLE
883:        (F:side camera) IDLE
884:        (IR:VidStop ) play 1 time(s) [Emitter1]
885:        (IR:Vid 1 ) play 1 time(s) [Emitter1]
886:        (IR:Vid 2 ) play 1 time(s) [Emitter1]
887:        (IR:Vid 5 ) play 1 time(s) [Emitter1]
888:        (IR:VidEnter ) play 1 time(s) [Emitter1]
889:        (F:front dr camera) IDLE
890:      End
891:
892:    EVENT: side camera
893:      If
894:        (F:side camera) is SET
895:      Then
896:        (F:back camera) IDLE
897:        (F:front dr camera) IDLE
898:        (IR:VidStop ) play 1 time(s) [Emitter1]
899:        (IR:Vid 1 ) play 1 time(s) [Emitter1]
900:        (IR:Vid 2 ) play 1 time(s) [Emitter1]
901:        (IR:Vid 1 ) play 1 time(s) [Emitter1]
902:        (IR:VidEnter ) play 1 time(s) [Emitter1]
903:        (F:side camera) IDLE
904:      End
905:
906:    EVENT: movie lights on
907:      If
908:        IR Seq:' ' Received within 3 seconds
909:      Then
910:        X10: B-11  PRE-Set Level 52 %
911:      End
912:
913:    EVENT: movie lights off
914:      If
915:        IR Seq:' ' Received within 3 seconds
916:      Then
917:        X10: B-11  OFF
918:      End
919:
920:    EVENT: movie phone interrupt
921:      If
922:        (F:PowerOn) is SET
923:        and CO: Ring 1
924:      Then
925:        X10: B-11  PRE-Set Level 29 %
926:        DELAY 0:00:01
927:        X10: B-11  OFF
928:      End
```

Conclusion:

and H e a d a c h e s . . .   !!

I made a promise at the beginning of this book to share with you all my problems (just with this installation). Now that it's done I can only say I'm glad I wasn't doing this one for a paying customer. This job was fraught with issues, mainly hardware related.

If I didn't work so cheap for me, I'd probably have fired myself. But then again, I'm pretty understanding when it comes to me. It mostly wasn't my fault, and fortunately, I believe myself.

I really didn't have any major problems with the wiring stage, but mainly because I knew from experience that I needed to check everything ONE - LAST - TIME before drywall went up. Actually, the only issue was a single wire running through the ceiling that a plumber (of all people!) had cut.

Even though I was the last person to do any significant work as far as mechanicals were concerned, I knew that contractors occasionally return to haunt their work with some final touches. But this was probably the first time that a plumber had cut my stuff. Just glad I caught it.

Early on during the finish stage I bought an LCD96M keypad (I thought!) at an auction. Well, now - it LOOKED like an LCD96M... it was MADE by RCS (who makes the LCD96M... but it didn't say "Stargate" on it. After hours of trying to figure out why it worked intermittently, I was informed that it was for a Comstar system. Well, send it back (WHOOPS - can't. Bought it at auction)....

I hinted earlier about the problems I had with the Caddx panel and the download software. I went through two panels with corruptions of the master access AND download access codes. On the third panel I did all my programming from the keypad to avoid it happening again.

But this wasn't all. For the longest time, I couldn't seem to get Stargate and Caddx communicating over the serial connection..... but that had to wait because I developed hardware problems:

Somehow I wound up with problems on my Stargate (which is several years old). I'd noticed during the course of this installation that it began behaving oddly, no longer recognizing CID, getting occasional "GEW ..." errors on the history screen, etc.

Oh, and somewhere along the line I blew my power supply & had to replace it ....

After a discussion with Jeff Stein at JDS, I upgraded the software and firmware to 302e and tried to upload a new IVR voice library with a utility I downloaded from JDS. At this point I was worse off than when I'd started - I only succeeded in erasing what functional voice libraries I still had.

Reasoning that the voice chip was defective, Jeff kindly sent me another voice chip. This didn't help, either.

Reasoning this time that the main processor chip was defective, Jeff kindly sent me another main processor chip. Once again I used the same voice update utility to try to upload the voice libraries. Afterwards I discovered that I was supposed to use a DIFFERENT utility with this processor. It therefore became hopelessly corrupted.

Since my IVR board was an older model and parts were very scarce I then had to pack up the board and ship it off to JDS. A couple of weeks later it came back, repaired finally.

Well, now - back to the Caddx to SG communication problem. I tried everything I could think of: swapping cables, trying new DB9 to RJ11 connectors - I even hacked into a DB9 connector & rigged up a temporary setup with a jack so I could try different pinouts. I checked for voltage on the P0003 connector (just to make sure the Caddx board was OK) & it checked out fine....

And then - finally - I narrowed it down to a bad data (straight-thru 6-pin) cable between the two pieces of hardware. It was unbelievable. Simple - except I had already checked that several times before.

Anyway, it works now. The best I can figure is that there was at least one problem in addition to the cable, and somewhere along the line I unwittingly fixed it. Sometime stuff just happens.

You'll see. Probably.

So it comes time for me to wrap this up and wish you the best on your own project. I'm encouraged to think that you read this from beginning to end and found it helpful. At least I'd like to believe that you can see that anyone can do quite a bit on a modest budget. I'd say that (aside from audio/video equipment) I probably have just over 2 grand in the all the equipment in this book.

I realize that's not peanuts to most of us. But it's a far cry from the 10 to 30 thousand you *could* spend. The downside, of course, is the labor. It can be time-consuming to get everything just right. However, if it's a hobby - and if it's your house - and if it's something you get off on ... well, it's not exactly like work, is it?

So if you haven't begun your own project yet, I smilingly (*sic?*) invite you to do so. In all seriousness, you'll never be bored. And you'll never run out of new possibilities as you and your system become intimately familiar.

In fact, you really will have to be careful that your spouse doesn't become jealous.

Best of luck,
Andy Jackson

www.Integratorpro.com